Bayonets and Blobsticks

The Canadian Experience of Close Combat 1915-1918

Aaron Taylor Miedema

Legacy Books Press

Published by Legacy Books Press
RPO Princess, Box 21031
445 Princess Street
Kingston, Ontario, K7L 5P5
Canada

www.legacybookspress.com

The scanning, uploading, and/or distribution of this book via the Internet or any other means without the permission of the publisher is illegal and punishable by law.

All illustrations are used under fair use or with the consent of the copyright holder. If you believe your copyright has been violated, please contact Legacy Books Press.

This edition first published in 2011 by Legacy Books Press
1

© 2011 Aaron Taylor Miedema, all rights reserved.

Aaron Taylor Miedema
    Bayonets and Blobsticks: The Canadian Experience of Close Combat 1915-1918
    Includes index
    ISBN: 978-0-9784652-9-2
    1. History : Military - World War I 2. History : Canada - Post-Confederation (1867-)
    3. History : Military - Canada

Printed and bound in the United States of America and Great Britain.

This book is typeset in a Times New Roman 11-point font.

Rolling this book up and attempting to bayonet Germans with it will void warranty.

For the men who had to undergo the negotiation of close combat – it is to these men that both I and this work are indebted.

Table of Contents

Acknowledgments. iii

List of Figures. v

List of Maps. vi

Introduction. 1

Chapter I: Myths and Misconceptions. 7
    I. Casualty Statistics
    II. Use and Abuse
    III. Alternative Weapons
    IV. Moral Revulsion
    V. Bayonet in Battle
    VI. An Alternative Interpretation

Chapter II: 1870 to 1914, The Bayonet Before the War. 27
    I. European infantry tactics 1870-1914: Paradigm or balance?
    II. 1871 to 1899: From Practice to Theory
    III. 1899 to 1914: From Theory to Practice
    IV. Training and techniques of British bayonet fighting 1849-1914
    V. Conclusion

Chapter III: Fear and Function. 43
    I. Close Combat: Instinct and Training
    II. Close Combat Conditioning
    III. The Negotiation of Close Combat

Chapter IV: 1915, The Bayonet and Trench Warfare. 59
    I. The Challenge of Trench Warfare
    II. From Canada to France
    III. Ypres: Baptism of Fire
    IV. Festubert: The Problem of the Attack
    V. Givenchy: New Tactics and Unforeseen Consequences
    VI. Bayonet Training 1915: Formal and Informal
    VI. Conclusion

Chapter V: 1916, The Bayonet and the Battle of Materiel. 86
    I. The Problem of the Offensive
    II. St. Eloi Craters and Mount Sorrel: New Divisions, New Corps, New Problems
    III. Interlude: The Lessons of the Somme
    IV. Courcelette: Tactics and Technology Old and New
    V. The Ancre Heights: Problems Old and New
    VI. Bayonet Training 1916: Organization and Control

Chapter VI: 1917, The Bayonet and the Set-Piece Battle. 109
    I. Vimy Ridge: Change and Continuity
    II. The Arras Offensive: Change Continues
    III. Hill 70: Charge and Counter-Charge
    IV. Passchendaele: Individual and Independence
    V. Bayonet Training 1917: Bayonet Fighting Matures
    VII. Conclusion

Chapter VII: 1918, The Bayonet and the War of Movement 134
    I. Amiens: The Bayonet and Open Warfare
    II. Scarpe: The Anatomy of a Bayonet Fight
    III. Arras: Continuing Patterns
    IV. Conclusion: Beyond the Canal du Nord

Conclusion. 148

Abbreviations. 152

Endnotes. 153

Works Cited. 200

Index. 217

Acknowledgments

THERE ARE A number of people who deserve acknowledgment and credit for the merits of this work. Foremost is Dr. Tim Cook, without whom this almost four year project would never have begun, and without whose assistance certainly it would never have come to completion. The second is Dr. Doug Delaney, who kept me focused on the larger picture when I became mired in the mountains of small details. Third, many thanks to Dr. Huw Osborne who has helped me to organize my thinking and to clarify my writing by his generous assistance in the revision of this text, and doubly so for doing it twice. The same thanks go out to Robert Marks of Legacy Books Press, who also went through the manuscript twice and helped to make it what it is, and additional thanks for putting up with my sometimes mercurial temperament and publishing it. Thanks also go out to the many people who have read the work and given me feedback and criticism. Finally, I express my gratitude to the long list of people who stood by patiently while I have expounded upon my ideas and who treated me with the appropriate amount of skepticism when I've said "just another couple of weeks and it will be done." For the failings of the present work I am myself responsible. I have had no end of good counsel from all the aforementioned people; it is my fault alone if I failed to take it or lost track of it. I must also give thanks to the Social Sciences and Humanities Research Council of Canada and the Joseph Armand Bombardier scholarship for their funding and support of this work.

Aaron Taylor Miedema
August 12, 2011
Kingston, Ontario

List of Figures

Figure 1: Bairnsfather cartoon illustrating the contrast between manning the trenches and the intense periods of close combat. 12
Figure 2: Bairnsfather cartoon showing a bayonet being used to dry socks 15
Figure 3: "Reaching for knife in contact after parrying the attack" and "Stab at groin" (*Bayonet Fighting Illustrated 1917*). 19
Figure 4: "The long point" (*Bayonet Fighting 1916*). 40
Figure 5: "Shorten Arms" (*Bayonet Fighting 1916*). 41
Figure 6: Men of the 29th Battalion practice the art of Bayonet fighting with wooden pugil sticks. 63
Figure 7: Plan for grenadier training. 64
Figure 8: "The jab" (*Bayonet Fighting 1916*). 83
Figure 9: The blobstick in use (*Bayonet Fighting 1916*). 106

List of Maps

Map 1: Canadian counter attacks, 22/23 April 1915. 65
Map 2: Battle of Festubert, 18 May 1915. 69
Map 3: Battle of Festubert, 20 May 1915. 71
Map 4: Battle of Festubert, 21-24 May 1915. 73
Map 5: Battle of Givenchy, 15 June 1915. 76
Map 6: The Canadian attack at Mount Sorrel, 13 June 1916. 90
Map 7: Battle of Courcelette, 15 September 1916. 96
Map 8: Attack on Regina Trench, 1 October 1916. 100
Map 9: Attack on Regina Trench, 8 October 1916. 102
Map 10: The Battle of Vimy Ridge, 9 April 1917. 113
Map 11: The Battle of Hill 70, 15 August 1917. 118
Map 12: Nabob and Nun's Alleys. 123
Map 13: Battle of Passchendaele, 26 October to 6 November 1917. ... 126
Map 14: The advance of the 116th, 42nd, 102nd, and 78th Battalions during the Battle of Amiens, 8 August 1918. 136
Map 15: The advance of the 7th and 8th Brigades during the Battle of Scarpe, 26 August 1918. 141
Map 16: 13th and 14th Battalion during the Battle of Arras, 2 September 1918. 145

Introduction

THE BAYONET IS one of the most iconic weapons of the Great War. The order to "fix bayonets" preceding the wait to before going "over the top" into No Man's Land has been immortalized in popular culture. However, equally prevalent is the deep-seated belief that the bayonet was a "pointless"[1] weapon against the new technologies of the late nineteenth and early twentieth centuries, such as magazine rifles, quick firing artillery, and machine guns.[2] This popular conception has led to the propagation of a myth that the continuing adherence to the bayonet demonstrated primitive doctrine by generals who were out-of-touch with the realities of modern warfare.[3] This notion of the bayonet's obsolescence fits with the popular conception of the Great War in literature, drama and film of terrible and senseless slaughter. Yet soldiers continued to "fix bayonets" before going into battle.

This contradiction led many historians to conclude that the bayonet was simply an irrational and conservative military tradition.[4] Some have called the adherence to the bayonet a "fetish,"[5] others have associated the bayonet with a "cult"[6] and given it a pseudo-religious significance. The bayonet had been a primary infantry weapon for more than two centuries and many have assumed that routine and ritual made soldiers fix primitive knives to the end of modern magazine rifles.[7] The idea of routine and ritual has also been used to explain the massive casualties on the Western Front as conservative generals threw waves of men against the new weapons technology in futile bayonet charges.[8]

The frequency of references to hand-to-hand fighting in official records and battalion histories demonstrates otherwise. Bayonet fighting and close combat also appear frequently in the personal accounts of soldiers. This suggests a vast difference in opinion between historians and the men who fought the war. Furthermore, the British Army increased bayonet training during the war. Between 1914 and 1916, while the overall training period for new British Imperial recruits dropped from 24 weeks to 14 weeks, the amount of instruction in bayonet fighting increased three-fold, from twenty to sixty hours.[9] Yet, in spite of this and other evidence, many historians still claim the bayonet was almost never used.

This book will examine the contradictions between wartime use and post-war historical conception of the place of the bayonet within rapidly-expanding arsenals and changing tactics of the Great War. To do this, this book will focus on the experience of close combat and bayonet fighting in the Canadian Expeditionary Force (CEF). This will illustrate the development and importance of shock tactics in the trenches of the Western Front and the role of the bayonet in the development of all-arms cooperation and the set-piece battle. In other words, this book will prove that, in spite of historical derision, the bayonet was one of the more important weapon systems of the war, and was one of the foundations on which doctrinal and technological change were built.

This investigation will begin by laying out the historiography of the bayonet in the Great War. The first chapter examines the myth of the bayonet's obsolescence and the arguments used to support this claim, summarizing and analyzing the criticisms of the bayonet, and the evidence upon which they were founded.

The second chapter presents the larger contexts that affected the role of the bayonet in European tactics. These include the intellectual and political factors governing the foundations of pre-war tactical thinking from the Franco-Prussian War (1870-1) to the outbreak of the Great War (1914), as well as the changes and refinement of the techniques and training of bayonet fighting in British Imperial forces from the middle of the nineteenth century to 1914. This exploration will demonstrate the period of intense debate and change in military institutions during the late nineteenth century and early twentieth centuries. This chapter focuses on European doctrinal theories and British bayonet practices, rather than on specifically Canadian sources. There are three reasons for this. The first is the relative dearth of pre-war Canadian sources on bayonet fighting.* The second is that Canada was engaged in fighting a European War as part of the British Expeditionary Force (BEF) and with British doctrine. The third is that the Canadian militia had agreed to meet the conditions of the Committee on Imperial Defense in 1911 and was, at the very least, coming into line with British practice by 1914.[10] Any lingering vestiges of domestic training and theory had been done away with when the British took responsibility for the training of the First Canadian Contingent in October of 1914.[11]

The third chapter examines the nature of close combat. This will first focus on how the close proximity of the enemy affects the individual combatant, and how it created a tension between the instinct for self-preservation and the use of training to condition soldiers to overcome resistance and engage the enemy at close quarters. Shock tactics capitalized upon this tension, as soldiers determined whether or not close combat was to occur. The function of the bayonet went far beyond thrusting it into an enemy soldier – it emboldened soldiers and forced the enemy to surrender or put them to flight. Examining this will clarify the logic of the historically-derided concept of *élan* and explore the "cult" – or more accurately the "culture" – of the bayonet.

The fourth chapter examines the first eight months of the war and the onset of trench warfare, following the experiences of the 1st Canadian Division from the outbreak of the war to the attack in 1915 at Givenchy. This will include the role of the bayonet in Canadian offensive operations during the battles of 2nd Ypres, Festubert, and Givenchy – during which, despite the complexities of bringing attacking infantry into bayonet-charging range, the bayonet proved a capable weapon when infantry could be brought into contact with the enemy. To bring the infantry into bayonet range, the Canadian infantry battalion began the process of assimilating a host of new weapons, including hand grenades, rifle grenades, Lewis guns, and trench mortars. These new weapons and tactics brought additional complexities to the battlefield, but they were all harnessed to permit the infantry to close to within bayonet-charging distance of the enemy. The chapter will conclude with the state of bayonet training in 1915 and the varied responses to the problems posed by mobilization and trench warfare,

---

* The only extant source available regarding Canadian bayonet training between 1900 and 1914 is found in: Department of Militia and Defence, *Infantry Training 1904* (Government Printing Bureau, 1904), part II, section iii. It contains three paragraphs that cover the proper assembly of troops to practice a single technique of bayonet fighting, that of the "thrust." Additional information may have been contained in *Infantry Training 1913*, which served as the training manual governing the training at Valcartier in August and September 1914; however, it seems this manual is no longer extant. See daily orders for Valcartier camp: LAC RG 9 II, F-9, vol. 1702.

such as the development of informal training and the appearance of unofficial techniques, and the establishment of a bayonet training administration to control informal and unofficial practices.

Chapter Five continues to trace the challenges of offensive operations on the Western Front and attempts to bring the infantry into bayonet range. The increasing weight of artillery supporting the advance across No Man's Land led to the refinement of tactics during 1916 and improved the chances of closing to within bayonet range; however, even after two years experience, getting the infantry to this range was still problematic. The combination of trenches, barbed wire, and the machine guns proved resilient in the defense. The Canadian attacks at Mount Sorrel, Courcelette, and Regina Trench illustrate the interplay between changing tactics, technology, success, and the bayonet. The chapter will conclude with the chain of command's efforts to bring uniformity to bayonet techniques and training.

The sixth chapter focuses on the Canadian Corps successes and tactical changes during 1917. The Canadian attacks at Vimy, Hill 70, and Passchendaele all demonstrated refinements in tactics and technology as well as continuity with prewar concepts of battle. Bayonet fighting itself underwent the same process of tactical refinement as other elements of the attack in 1917, evolving as a result of practical experience.

The last chapter examines the bayonet in offensive operations of the Canadian Corps during "the last hundred days" in the fighting at Amiens, Scarpe, and Arras. In the final months of the war, trench warfare gave way to open warfare and the nature of operations changed significantly. However, even in light of the tactical and technological developments over the course of the war, the bayonet still figured prominently in Canadian attacks.

From this investigation of Canadian offensive operations, two main points can be proven. The first is that the bayonet remained prominent in Canadian attack doctrine throughout the war. There were substantial changes in the weapons and tactics used in the attack between 1914 and 1918, but they served the same objective: to bring the infantry into close grips with the enemy and silencing resistance by forcing them to fight, flee, or surrender. Canadian operations demonstrated that an increasingly-sophisticated combination of artillery, machine guns, and new tanks supported the infantrymen as they closed on enemy positions with the bayonet. Tangible changes occurred in tactics and technology during the war, but these changes were assimilated into a contiguous larger conception of the infantry attack that maintained an important role for the bayonet. Thus, the bayonet was as integral to Canadian operations in 1918 as it had been in 1915.

Second, bayonet fighting was part of the process of refining weapons and tactics over the course of the war. The training and techniques of bayonet fighting from 1849 to 1918 demonstrates that bayonet fighting was not static.[12] For example, the British theory, analysis, and techniques of bayonet fighting between 1902 and 1914 reflected institutional attempts to reshape bayonet fighting to incorporate the lessons of the Boer and Russo-Japanese Wars. Also, an administration to control bayonet training was established in 1916. This was a response to the strains put on the training systems by mass mobilization and as well as the unofficial techniques and informal training systems front-line soldiers had developed. Finally, the refinements of bayonet fighting techniques, in the form of close quarters combat known as *"infighting"* and unarmed combat, demonstrates the maturity of a martial art refined through experience. Far from a static anachronism, bayonet fighting was constantly being refined to meet the problems of the Western Front and modern warfare.

Both these conclusions rely upon the assertion that the bayonet was frequently used on the Western Front. There are a substantial number of references to challenge the notion of the bayonet's obsolescence. Tim Cook wrote that "too many references to successful bayonet use are made in after-battle reports and in the account of medal citations to discount the effectiveness of the weapon in battle."[13] A survey of the 58 Victoria Crosses awarded to members of Canadian infantry formations supports Cook's conclusion: 13 of these Victoria Cross citations include direct reference to bayonet use or hand-to-hand fighting.[14] The accounts of bayonet fighting in official records, battalion histories, and personal accounts of the Canadian Corps number in the hundreds.

Bayonet fighting was frequently taken for granted by the soldiers writing operational reports. Individual recollections of bayonet fighting were often obscured in official records by the use of such terms as "cleared," "secured," and "opposition." For instance, in addition to the 13 definitive instances of bayonet fighting and hand-to-hand fighting in the 58 Canadian infantry citations for the Victoria Cross, there are an additional 20 citations that use language that could imply bayonet use ("charged," "stormed," and "rushed").[15] Personal accounts provide a more detailed record of the highly-personal task of close combat and help to clarify official records; however, they fail to offer a complete picture of bayonet fighting. Some veterans were unwilling to talk about the experience of close combat and others were not capable of recollecting the experience. For example, when asked in a 1960s CBC interview whether he used his bayonet at Kitcheners Wood in April 1915, Charles Lunn answered "Oh definitely, definitely...."[16] He then commented on the pressure faced by veterans when discussing close combat: "it's just one of those things you didn't talk about."[17] Lunn felt no such pressure, but his comment raises the question of how many accounts of bayonet fighting went unrecorded because killing in hand-to-hand combat was "just one of those things you didn't talk about." Official records often omitted accounts of bayonet fighting, but Lunn's comment suggests that soldiers did as well.* Of course a soldier also had to survive in order to leave a personal record of close combat, and many accounts of bayonet fighting have been lost amongst the lists of the fallen on the Western Front. Nonetheless, the instances that were noted in official records, battalion histories, and personal accounts provide a significant challenge to the argument that the bayonet was rarely used in the Great War.

---

* An example of the emotional strain felt by some soldiers about close combat, even 50 years afterwards is found in the interviews conducted for the CBC for *In Flanders Fields*. LAC RG 41, vol., 16, 102nd Battalion, Roy Gross, 1/12-13.

"GROSS. And all at once the noise around you, bullets flying, but nothing was hitting you, so I looked around, and there was my partners, they were being mowed down. And I couldn't make out – and in the meantime there was quite a few Germans prisoners going by, and we would just say 'back,' and they would go back on their own. And look around, and there was a little Prussian officer there, and he had an automatic pistol, and he was shooting. You know, although he was a prisoner. I wouldn't – I wasn't brave but I was mad, and the only thing – I made one rush at him, and since I thought oh, my goodness, if he had one shot, he would mow me down good job he never had another shot. Mind you I was no hero when I done it. But I just hopped on the jump and I just clean knocked him down. And there was no more Prussian guard, but –

INTERVIEWER. Did you have to use a bayonet on him?
GROSS. I won't say.
INTERVIEWER. Why not. Well – you are telling me all about the war except the part – that is war, that is killing.
GROSS. I didn't know what I was doing.
INTERVIEWER. Yes, you did. You had to stop him, didn't you.
GROSS. I had to stop him.
INTERVIEWER. All right, so you had to kill or be killed. Wasn't it?
GROSS. I up with my Kitchener boot, and kicked him under the chin, and I think that is what done it, he was no more.

Chapter I: Myths and Misconceptions

THE POPULAR NARRATIVE, which saw soldiers as hapless pawns slaughtered in battles planned by out-of-touch generals, lies at the heart of the criticisms of the bayonet.[1] This interpretation has become known by the title of "lions and donkeys." The brave infantrymen were cast as the lions and the generals who squandered their lives were cast as the donkeys. This narrative originated in the 1920s and 1930s, as soldiers and politicians began writing about their experiences of the war. These accounts reflected a broad spectrum of attitudes from patriotism to disillusionment.[2] However, as the optimism of the postwar world faded the work of some disillusioned veterans, and the themes of futility and waste, caught the popular imagination. These works also began to shift blame between political and military leaders for mistakes and errors made in the conduct of the war.[3]

The rise of fascism and the Second World War came to dominate popular culture and historical writing during the 1940s and 1950s, but as the fiftieth anniversary of the outbreak of the Great War loomed, there was a resurgent interest. With the official archives still closed to the public, historians were forced to rely on the literature written by soldiers and politicians in the 1920s and 1930s, inheriting the themes of futility and blame. These themes were magnified by the experience of the Second World War, which, due to the horrors of the defeated fascist regimes, was seen as "a good war."[4] The Great War in contrast was seen as a futile conflict that seemed to have accomplished little, and in some ways set the stage for the Second World War. Guided by the themes of futility and waste British historical attention became fixated on the high casualty rates of the first day of the Somme and to a lesser degree at Passchendaele.[5] The result was the creation of a narrative with "brave helpless soldiers disillusioned by blundering obstinate generals who achieved nothing in return for systemic slaughter."[6] British historical writing throughout the 1960s reiterated this hostile attitude towards the generals.[7]

The "lions and donkeys" narrative was given its formative shape and name by Alan Clark in *The Donkeys* (1961) and gained significant popularity throughout the 1960s. This hostility presented a digestible version of the war for the public, complete with heroes and villains. This interpretation was not only prevalent in historical writing, but also in the arts. For instance, the play, and later the film *Oh What a Lovely War!* (1963 and 1969) upheld and propagated the "lions and donkeys" interpretation in popular culture. Twenty years later, *Blackadder Goes Forth* (1989) demonstrated the lasting popular narrative, an interpretation that has continued to influence historical writing and popular culture – British and otherwise – for almost a half century.[8]

The bayonet featured prominently in the "lions and donkeys" narrative, as it supposedly demonstrated how out-of-touch generals were with the realities of the Western Front. These generals continued to throw masses of men against machine guns in futile bayonet charges because, it has been presumed, battles had always been fought in this manner. John Ellis in *Eye Deep in Hell* (1976) made one such association: "They [the generals] revered the bayonet as the *arme blanche* and a more than adequate response to anything produced by the technological revolution."[9] John Harris, writing of the training of British infantry for the Somme, also used the bayonet as proof of out-of-touch generals:

> Their officers had been inculcated with the belief that they must obey orders without question and the men had been taught to rely chiefly on the bayonet for killing: even when they arrived in France they were still kept at bayonet practice although experience had long since proved that bombs were far more efficacious. For the most part they had no knowledge whatsoever of trench fighting.[10]

As a result, the bayonet had become one of the critical pillars on which the "lions and donkeys" narrative rested.

Although historians such as John Terraine have been challenging the "lions and donkeys" narrative since the 1960s, another challenge arose in the discussion of doctrine and technology in the Great War in the early

1990s. Bill Rawling's *Surviving Trench Warfare* (1992) examined the introduction of new weapons, tactics, and technologies into the Canadian Corps over the course of the war and revealed a culture of experimentation and development within the Corps. Rawling also demonstrated that generals, rather than being donkeys, were actively experimenting with new ideas in order to break the deadlock. This was followed by Paddy Griffith's *Battle Tactics of the Western Front* (1994), which examined the factors governing the evolution of British attack doctrine during the war. This new narrative – coined later as "the learning curve" – has provided a far more detailed appreciation of the war and in some measure helped to further redeem the generals – in academic circles at least – from their reputation as donkeys.[11]

The impact of the "lions and donkeys" narrative runs deep, even in academic history written in Canada. However, within Canadian historiography the themes of futility and waste are not as prevalent. Tim Cook observed that in Canadian historical and popular memory the high casualties during the war were "balanced" by the operational record of the Canadian Corps and the popular narrative as a fundamental step in the development of Canadian nationhood.[12] However, when the bayonet is examined in Canadian histories of the Great War, the influence of the "lions and donkeys" interpretation is clearly evident. One such example appears in Jonathan F. Vance's *Death So Noble* (1997), in his investigation of myth in Canadian historical accounts of the Great War:

> Unit histories are no more critical. They have little to say about the political or strategic decisions that killed off so many of their members, and they rarely question the ability of Allied generals to carry out their duties. Repeated bayonet charges against well-defended trenches demonstrate the tenacity of the infantryman, not the tactical bankruptcy of his commanders.[13]

Although Vance was referring to the lack of critical analysis found in battalion histories, he implied that generals were tactically bankrupt because of their adherence to the bayonet. In Canadian historical writing, Vance is not alone in propagating of the myth of incompetent generals squandering lives in bayonet charges.[14] To be fair, the primary intention of these historians was not to examine bayonet fighting and these are merely peripheral remarks, but the frequent repetition of the bayonet as the hallmark of "tactical bankruptcy" has erroneously entrenched the myth of the bayonet's obsolescence.

Even close examination of the bayonet has not been enough to overcome this myth. Tim Ripley's *Bayonet Battle* (1999) provided an example of the persistence of the "lions and donkeys" interpretation in spite his focus on the bayonet. Ripley's work focused on the bayonet in the twentieth century and concluded that "over the next hundred years the order to 'fix bayonets' will still reverberate around the battlefield when the going gets tough."[15] The Great War figured prominently in Ripley's investigation of the bayonet and he included a sizable chapter devoted to reviewing instances – successful and otherwise – of bayonet charges and hand-to-hand fighting during the Great War. Yet, in spite of Ripley's intentions and wealth of operational evidence, he was still compelled to conclude that hand-to-hand fighting in the Great War occurred only on "rare occasions."[16] Furthermore, he contradicted himself, echoing the statement of Vance, in his description of British tactics during the war: "The bayonet charges of the British infantry clearly demonstrated the bravery and discipline of the Tommies, but also their generals' totally fool hearty tactics."[17] Neither Vance nor Ripley provided citations for the origin of this conclusion; this idea had simply been assumed to be true.

Often it is the case that the assumption of the bayonet's obsolescence greatly influenced the interpretation of the evidence. For example, Paul Hodges's article "They don't like it up 'em" observed the "fetishization" of the bayonet during the Great War and linked it to atrocities by British soldiers against surrendering Germans. One of the most prevalent notions of this fetish was the constant repetition of the idea that the enemy could not stand up to a bayonet charge by British soldiers; rather, they would surrender or flee. Another was associating bayonet training with a combination of hatred, dehumanization of the enemy, appeals to masculinity, and even sexual imagery.[18] He also argued that this "fetishization" of the bayonet developed a life of its own and was completely isolated from battlefield reality. An isolation caused by Hodges assumption that "the bayonet was not a particularly useful or effective weapon."[19] The result, according to Hodges, was that British soldiers were unable to satisfy this fetish in battle and therefore fed the fetish by executing surrendering Germans.[20] In reality, this "fetish" had a practical purpose, it helped to trivialize the task of bayonet fighting and motivated soldiers to kill the enemy in close quarters, indeed this included prisoners, but, it also included combatants. This conditioning also involved the engendering hatred of the enemy; Hodges cited the 1917 musings of British infantry Captain Stormont Gibbs in order to make this conclusion:

> In any sort of hand fighting there are the savage emotions that motivate the shot or thrust. The great horror of war is this prostitution of civilized man. He has to fight for his country and to do so has in actual practice to be brutalized for his country; he has to learn to hate with the primitive blood lust of the savage if he is to push a bayonet into another human being (who probably no

more wants to fight than he does). Need he hate? In the case of the average man he must as the counter-balance to fear.[21]

Gibbs acknowledged the presence of conditioning ("hate") in order to overcome instincts of self-preservation ("fear"). However, Gibbs also contradicted Hodges's claim of the bayonet's ineffectiveness with the suggestion that conditioning was necessary specifically because of "hand[-to-hand] fighting."

While the peripheral comments made by historians have compounded the myth of the bayonet's obsolescence, the evidence to support their claims is scant. Historians who have provided evidence for the bayonet's obsolescence typically cite one or several of the following arguments: First, the low numbers of men recorded as wounded by bayonet;[22] second, the misuse of the weapon by soldiers;[23] third, the use of alternative close combat weapons;[24] fourth, an inherent moral revulsion to killing at close quarters; and finally, the perception of the one-sided fight between a soldier armed with a machine gun and the soldier armed with a bayonet.[25] However, when investigated point by point, each of these supporting arguments prove flimsy at best.

I. Casualty Statistics

The most frequent argument made to support the myth of the bayonet's obsolescence is the low ratio of bayonet wounds found in medical survey statistics. British and Canadian historians have primarily relied on the findings of T. J. Mitchell's and G. M. Smith's, *Medical Services: Casualties and Medical Statistics of the Great War,* which determined that of 212,659 cases only 0.32 percent of non-fatal wounds were inflicted by bayonets.[26] A few have relied on the American study by Harry L. Gilchrist, *A Comparative Study of World War Casualties: from Gas and Other Weapons*, which placed the ratio of bayonet wounds even lower at roughly 0.1 percent of 224,089 cases.[27] Others have used smaller surveys in arguing the obsolescence of the bayonet. For example, Tom Henderson McGuffie cited a survey of 698 non-fatal eye wounds from the war, of which the bayonet caused only one, or roughly 0.14 percent.[28] As a result of these statistics, several historians have claimed that bayonet fighting was infrequent. However, wounds delivered by the bayonet reflect only a portion of the bayonet's function and impact on the Western Front.

Figure 1: Bairnsfather cartoon illustrating the contrast between manning the trenches and the intense periods of close combat.

While surveys provide historians with concrete numbers and wide samplings, there are several problems with the direct correlation between the number of men non fatally wounded with the bayonet and the frequency of bayonet use on the Western Front – the first being that these statistics are strictly those of men non fatally wounded by the bayonet. There were other important categories of casualties that are related to bayonet use: for instance, the number of enemy killed or taken prisoner by the bayonet. This leaves room for the possibility that bayonet wounds were a statistical anomaly; therefore, permitting the alternative conclusion that the bayonet had a greater capacity to kill or intimidate than to wound. Another problem with wound statistics is that they fail to take into account the complex and cooperative effect of weapons in the Great War: many German soldiers fled in the face of the bayonet charge only to become casualties by artillery or small arms fire. In these cases, the bayonet did not inflict a physical wound but played a role in inflicting the casualty. While these overlooked aspects of the bayonet may be impossible to quantify, it is certain they added to bayonet-related casualties.

The number of soldiers killed by the bayonet, for which there are no complete statistics, has led some historians to argue the bayonet was a particularly lethal weapon, more frequently killing than wounding.[29] Soldiers were trained to thrust for lethal areas of their opponent's body: the head, neck, and chest.[30] These target areas were meant to incapacitate their opponent quickly and killing their opponent outright equally served the purpose. The idea the bayonet was more likely to result in death than injury is also supported by the dearth of soldiers wounded in hand-to-hand fighting or losing a bayonet fight with the enemy, in the course of this investigation only a single Canadian instance was found.* This point was once again echoed in the training manuals: "In a bayonet assault all ranks go forward to kill or be killed, and only those who have developed skill and strength by constant training will be able to kill."[31] There are few statistics for bayonet fatalities, but the few that do exist illustrate a vast discrepancy with the wounding surveys. For instance, in the official records of the Canadian Corps, there appear two particular occasions where kill ratios with the bayonet in battle were recorded, and both indicate this ratio as having approached or exceeded 50% of all enemy fatalities.[32] This suggests that in

---

* This instance was the bayonet fighting on September 9th, 1916 at Poziers, for which Leo Clarke was awarded the Victoria Cross. Clarke was wounded in the knee during the course of the fighting. W. W. Murray, *The History of the 2nd Canadian Battalion (East. Ontario Regiment) Canadian Expeditionary Force in the Great War 1914-1919*, (Ottawa: 1947), 126-7 and <www.thememoryproject.com/digital-archive/profile.cfm?cnf=cf&collectionid=987>.

close combat during the Great War there were two typical outcomes: unscathed or dead.

In addition, fatalities and wounds were not the only measure of the bayonet's impact on the battlefield; the bayonet frequently intimidated defending soldiers into surrendering or fleeing in the face of a determined attack. The bayonet charge forced defending troops to decide whether to fight, flee, or surrender. The negotiation of surrender often took place inside the range at which the bayonet charge had been launched and a substantial portion of the more than 42,000 German soldiers taken prisoner by the Canadian Corps also needs to be considered in assessing the role of the bayonet on the Western Front.[33] The bayonet charge could also rout defending troops, at which point, as mentioned above, some became victims of artillery or small arms fire. The War Diary of the 85th Battalion recorded a minor attack in the Souchez sector in late June 1917 and observed the fate of many troops who chose not to engage the attacking Canadians in close combat: "Both objectives were taken - 4 machine guns were captured - 40 or 50 enemy killed in hand-to-hand fighting in [the] area occupied by us besides those caught, while retreating, by our fire."[34] Certainly the bayonet played a role in causing these additional casualties.

Wounding statistics in general are deceptive, and obscure the impact of the bayonet. The bayonet, being a short-range weapon was only effective in an offensive context. Broad wounding surveys included substantial wastage inherent in manning the trenches and have, as a result, subsumed the significance of the bayonet in the attack. In battle the ratio of bayonet wounds, although not quantifiable, was undoubtedly higher. Critics of the bayonet have frequently cited the low ratios offered by postwar medical surveys – however, these surveys presented a distorted picture of the casualties inflicted by bayonet. The number of soldiers that were captured, routed, or killed by the bayonet are not considered by critics of the bayonet who have based their conclusions on these surveys. Wounding statistics, while providing tangible numbers, present an incomplete picture of the role of the bayonet in the Great War, and thus are too incomplete to be an accurate indicator of the frequency of bayonet use.

II. Use and Abuse

Noted "lions and donkeys" historian Denis Winter summed up the bayonet in *Death's Men* (1979) in the following terms: "for most of the Great War it was simply an anachronism, useful as a toasting fork, biscuit slicker or intimidator of prisoners."[35] Tom McGuffie gave a similar review: "The bayonet's only really useful role was to open a tin, chop fire wood, prise open a door or hold bread or meat to be toasted before a fire."[36] The illustrations of the popular British trench cartoonist, Bruce Bairnsfather, show the bayonet being used in all these ways, as well as for buttering bread and drying socks.

Figure 2: Bairnsfather cartoon showing a bayonet being used to dry socks.

Some of these misuses are found in the Canadian soldiers' discourse, giving credence to misconceptions of the value of the bayonet.* The 14th Battalion history provides one instance of a bayonet used as a can opener when the unit received a shipment of "Irish Butter":

---

* Ronald MacKinnon in a letter home made a complete list of answers in a letter home: "We were having a lecture on bayonet fighting this morning from a man who had never seen France and never heard a shot fired in anger. He asked the class what the bayonet was used for. Here are some of the answers he got: cutting cheese, chopping wood, toasting fork and killing rats. Imagine a man with a staff job who has never seen any fighting giving a lecture on bayonet fighting to men who have seen all the way from 2 years to 2 weeks in the trenches! " Ronald MacKinnon, 15 November 1916. From Canadian Letters & Images Project.

Eagerly the cans were rushed to the cooks, who were ordered to waste no time in putting the contents to use. Meanwhile, an individual greedier, or perhaps it would be charitable to say, hungrier, than the rest, was digging at the cover of his can with a Lee-Enfield bayonet. Soon the point penetrated and simultaneously visions of golden butter faded. From the tin there escaped, like soda water suddenly released, a sizzling fluid, foul smelling and horrible. 'If that's Irish butter', remarked the N. C. O. disappointedly, 'thank God we have no Irish cheese.'[37]

Several additional uses are found in the personal accounts of Canadian soldiers. Harold Peat, and others, found the bayonet a critical tool for trench cooking: "If we attempted to cook anything we would stick a bayonet into a sandbag and hang a brazier on it, then cook in our mess tins over that."[38] Canadian soldiers were also willing to turn their bayonets on rats, and the rat hunt is a common story in personal accounts.[39] A. G. Sinclair recalled one rat hunt in billets during the early months of 1915: "We had a great chase after a rat round the ceiling beams of our room yesterday. It succeeded in getting away, even though one enterprising youth went after it with his bayonet fixed."[40] Sometimes alternative uses were motivated by battlefield desperation. At 2nd Ypres, Harold Baldwin turned his bayonet into an entrenching tool in a pinch:

> I then felt a very earnest desire to live, and when the next halt came and the shells were coming over in a never-ending stream, I had an intense desire to explore the bowels of the earth. On feeling for my entrenching tool, to my dismay, I found it gone. Grabbing my bayonet from the scabbard I went to work, and the way I burrowed with my hands on that bayonet was a caution. I would not have been taken aback to see a prairie badger.[41]

Soldiers found numerous alternative uses for their bayonets on the battlefield, in the trenches, and in billets. However, none of these alternative uses of the bayonet provides credible evidence of the bayonet's value – or lack of it – as a weapon.

The exploration of the bayonet's alternative use serves primarily to demonstrate that soldiers were resourceful when coping with the environment of the Western Front. The weakness of the argument of alternative use is that it fails to draw any conclusion regarding the bayonet's merit as a weapon. Historians of the "lions and donkeys" narrative have attempted to imply that these alternative uses were all that the bayonet was good for. As well, alternative use of the bayonet served to heighten the contrast between the "donkeys," who revered the bayonet, and the "lions," who pragmatically used the bayonet as a tool. However, the indictment of the bayonet on the grounds it was a useful implement in soldiers' lives relies on the pre-existing assumption that the bayonet was obsolete in the Great War. In reality, the soldiers' improvisations demonstrate little but the resourcefulness of soldiers.

III. Alternative Weapons

The resourcefulness of soldiers during the Great War was also demonstrated by the proliferation of alternative close combat weapons. Soldiers independently sharpened spades, constructed trench clubs, and acquired knives to equip themselves for the rigors of close combat on the Western Front. One soldier was given the suggestion to carry a rusty fork in his puttee in order to stab Germans in the eye.[42] For critics, the use and acquisition of alternative weapons has been interpreted as resulting from an inadequacy of the bayonet. Typically, they argue that the bayonet was considered too long and unwieldy for fighting in trenches.[43] If swung widely, the bayonet could easily get snagged on a sandbag or embedded in the wall of a trench – however, a wide swing with the bayonet in the attack or defence was an indicator of poor training. The training literature advocated the use of techniques that kept the point of the bayonet in line with an opponent. The training literature also indicated that alternative weapons were secondary to the bayonet. Indeed, when examined closely, the use of alternative weapons suggests only that hand-to-hand engagements were a frequent occurrence on the Western Front.

The argument that alternative weapons were meant to supplant the bayonet at first seems a persuasive one, with numerous examples of alternative weapons found both in museums and in the pages of personal accounts of soldiers.[44] However, it fails to take into account that these improvised weapons were intended to be used with, not instead of, the bayonet. For instance, the equipment issued to trench raiders on the night of 16 December 1916 illustrates that the trench club – or "knob kerry" – and the bayonet were *both* key pieces of raiding equipment:

> The above parties will be equipped as follows...
> Bayonet Men - Rifle, *Bayonet*, Torch, 20 rds. S.A.A., 6 Mills No. 5...
> Wirer - Roll of wire, 2 Mills No. 5 and *Knob Kerry*...
> Gun Cotton Men. - 2 Mills No. 5, Torch, 10 lbs of Gun Cotton, Fuse for G. C., *Knob Kerry*...
> Conductor. - Torch, *Knob Kerry*, 2 Mills No. 5...[45]

In contrast to claims the alternative weapons were meant to replace the bayonet, the knob kerry was being used in conjunction with the bayonet by soldiers unable to carry a rifle as they were laden with heavy equipment or, who required their hands to be free: for the stringing of barbed wire, the setting of explosives, and directing traffic during a raid.[46] The ordering of the men in their respective trench storming parties during this raid was also significant:

> PARTY B. One Bayonet Man, One thrower, One Carrier, One Loader, and One wirer...
>
> PARTY C. One Bayonet Man, One Thrower, One Carrier, One Loader, One Rifle Man and One Wirer...
>
> PARTY D. One Bayonet Man, One Thrower, One Carrier, One Loader, One Rifle man and One Wirer...
>
> PARTY E. One Bayonet Man, One Thrower, One Carrier, One Bayonet Man, One Loader, and One Thrower...
>
> PARTY F. One Bayonet Man, One Thrower, One Carrier, One Loader, Two Gun Cotton Men...
>
> PARTY G. One Bayonet Man, One Thrower, One Carrier, One Rifle Man, One Thrower, One Carrier, One Loader, One Conductor...
>
> PARTY H. One Bayonet Man, One Thrower, One Carrier, One Loader, One Rifle Man, and One Wirer...
>
> PARTY I. One Bayonet Man, One Thrower, One Carrier, One Thrower, One Conductor, One Loader...[47]

In this particular raid, a single bayonet man led each party in the trenches, but frequently two or more bayonet men led the raiding parties.[48] Ormond St. Patrick with the 47th Battalion recalled the composition of his trench storming party in a raid on the evening September 16th, 1916:

> Well of course we had quite a bit of training before hand, I was a number two bomber you know we had a mills bomb there we had two fellows as bayonet men they went ahead you see in the trench. And there were two bombers and I think there were about may have been seven or eight of us and Keiller was in charge Lt. Keiller was in charge of the party our party.[49]

This was in accordance with the training given to bombers since the spring of 1915; bayonet men led the individual parties and were intended to deal with close combat situations that arose as these parties pushed down enemy trenches.⁵⁰ In the tight confines of a trench, with sharp corners between bays and numerous entrances from dugouts, every member of the raiding party needed a close combat weapon of some sort to deal with potential threats that could appear from anywhere.

Figure 3: "Reaching for knife in contact after parrying the attack" and "Stab at groin" (Bayonet Fighting Illustrated 1917).

Soldiers also privately acquired a wide variety of trench and puttee knives – shorter hand held blades or spikes, often featuring knuckle dusters and other offensive surfaces. The knife was considered a back-up weapon for troops engaged in hand-to-hand combat. The use of the knife had been advocated in official training literature in 1916.[51] The Canadian training pamphlet *Bayonet Training Illustrated 1917*, for instance, introduced techniques in which a puttee knife was thrust into the opponents groin after two bayonet fighters had come into direct contact.[52] Rather than supplanting the bayonet, these alternative weapons were integrated into operations and the techniques of bayonet fighting.

IV. Moral Revulsion

The argument that there is an inherent moral revulsion to bayonet use is an argument put forth in David Grossman's book *On Killing* (1995) and relies upon the use of alternative weapons. However, his argument differs somewhat from those based strictly on alternative weapons and needs to be addressed separately. Grossman focused on the psychology of combat and drew a number of conclusions regarding the role of the bayonet in the Great War. He, like many previous commentators, used wounding statistics to prop up his argument that "bayonet combat is extremely rare in military history;"[53] however, he concludes that the infrequency of bayonet fighting was due to the existence of an inherent moral resistance to stabbing another person with a thrusting weapon. Soldiers who willingly used the bayonet against another were, according to Grossman, of a malignant character. He supported his argument by observing numerous instances of German soldiers in the Great War preferring to use the butt of the rifle rather than the bayonet. Grossman also argued that the proliferation of alternative weapons, "clubs, coshes, and sharpened spades,"[54] was motivated by innate abhorrence to piercing the body of another human being with a sharp object. Grossman summarized his findings:

> Wound statistics from nearly two centuries of battle indicate that what is revealed here is a basic, profound, and universal insight into human nature. First the closer the soldier draws to his enemy the harder it is to kill him, until at bayonet range it becomes extremely difficult, and, second, the average human being has a strong resistance to piercing the body of another of his own kind with a handheld edged weapon, preferring to club or slash at the enemy.[55]

Grossman is correct in that distance is important in understanding shock tactics and bayonet fighting, but it is not a factor of morality.

Grossman's observation about the morals of slashing over thrusting and the tendency of German soldiers to use the butt of the rifle, observed what is an instinctual tendency. Grossman himself documented an example of this: "Prince Frederick Charles asked a [World War I] German infantryman why he did this. 'I don't know,' replied the soldier. 'When you get your dander up the thing turns round in your hand of itself."[56] In hand-to-hand fighting, with little or no training in close combat, the instinct is to swing across the body in order to ensure that an attack is gathered up and put aside. One sixteenth-century fencing instructor went so far as to rely on this instinctual response in teaching his students to ward an attack properly by using sharpened weapons in training.[57] *Bayonet Training 1916* also acknowledged the importance of defensive instincts and advocated aiming for the throat as it would often cause an opponent to instinctively flinch, or "funk," as the weapon came close to the opponent's eyes.[58] The increased pace of combat induced by decreased distance results in more strength and speed being applied to the swinging action and thus the actions become wider. With many close combat weapons, there was little option after a wide defensive swing but to swing back in order to attack or defend again. In the case of the bayonet, this instinctual sweep in defense often presented the butt of the rifle for a rapid attack. This provides a practical explanation of Grossman's observation why German soldiers, as well as many others, wielded their rifle butts instead of thrusting with the bayonet.

Grossman himself provides an example of a German lance corporal that both refutes his own arguments and highlights the importance of combative training in overcoming the instinct to swing wide:

> We got the order to storm a French position, strongly held by the enemy, and during the ensuing melee a French corporal suddenly stood before me, both our bayonets at the ready, he to kill me, I to kill him. Saber duels in Freiburg had taught me to be quicker than he and pushing his weapon aside I stabbed him through the chest. He dropped his rifle and fell, and the blood shot out of his mouth, I stood over him for a few seconds and then I gave him a coup de grace. After we had taken the enemy position, I felt giddy, my knees shook, and I was actually sick.[59]

Grossman used this example in an attempt to demonstrate a moral revulsion to stabbing another human being. However, the soldier's physical response to the engagement failed to suggest that the giddy feeling, shaking knees, and vomiting were anything but a response to the adrenaline induced by the stress of battle and close combat. What Grossman glosses over in the account is the German soldier's claim that his bayonet thrust owed much to his training in the sport of Schlager fencing. More significantly, combative training permitted this lance corporal to recall the fight clearly and to attack by binding his opponent's weapon and thrusting home with his bayonet in the same action. Fencing practice had allowed him to remain in cognitive control of the situation and his bayonet training permitted a proper defensive response with the rifle to ensure the bayonet was in the correct place to deliver the point.

Training in close combat could overcome, or at least reprogram, the instinct to swing in defense. Through repetitive drill, soldiers developed the instinctive responses that utilized the capabilities of the bayonet most effectively. Thus, even suffering from the daze induced by meeting the enemy in close combat, soldiers were still able to use the point, rather than the butt, against their opponents.[60] With even more training and sparring against an opponent, as seen in Grossman's example of the German lance corporal, soldiers could condition themselves to remain in control of their faculties in the stress of close combat. However, within the constraints of military training syllabi and the strains of mobilization for the Great War, offering soldiers these opportunities were limited. The German lance corporal above had received training in combative sports before the war and likely for several years; many soldiers were not so fortunate.

Inadequate training – of which there was no shortage in the Great War – resulted in soldiers reverting to untrained instincts when presented with close combat situations. For many, the rifle butt was simply the easiest form of attack after using a wide swing with the bayonet to ward off the opponent's attack. The rifle butt was taught as part of British systems of bayonet fighting – however, training literature made it clear that: "The butt should only be used when the attack with the bayonet has failed, and it is impossible to use the point."[61] Grossman's observation that soldiers adopted smaller weapons like clubs and sharpened spades is correct; however, there is little evidence that the adoption of these weapons was linked at all to a universal moral objection to thrusting. As discussed above, there were practical reasons for the adoption of alternative close combat weapons, not least of which being the frequent occurrence of close combat in cramped quarters during the Great War. Poor training provided another potential motivation for the acquisition of alternative close combat weapons. For poorly trained soldiers, these weapons were easier – being lighter and shorter – to recover after a wide swing across the body and thus better suited to untrained instincts within the close confines of a trench.

V. Bayonet in Battle

Another common criticism of the bayonet has been the folly of charging in tight-packed formations toward enemy trenches. This observation has been frequently used to argue the bayonet's obsolescence. The French attacks in the opening months of the war and British attacks on the first day of the Somme present examples of fairly close infantry formations advancing over hundreds of yards only to suffer crippling casualties. Critics of the bayonet argue that pushing men forward into bayonet range only led to horrendous casualties, and the continued emphasis on the bayonet during the Great War led to mass formations being driven toward machine guns. However, there are a limited number of examples of this from the war. In fact, the massed long-range bayonet charge was not the only manner of assaulting the enemy with the bayonet, nor was it the recommended manner. Pre-war British training literature had advocated a different approach to closing on enemy positions than an advance in tight formations. Furthermore, this literature stressed that the bayonet charge was not to be launched over hundreds of yards, but at thirty yards or less. To lay the blame for high casualties on the bayonet, as many critics have done, is far too simplistic and fails to observe the tactical and strategic complexities facing armies during the Great War.

Much of this criticism of the bayonet originates from the failure and high casualties of the French Plan XVII in the opening weeks of the war, which involved a vigorous general offensive against prepared German positions on the Franco-German frontier. These attacks were often portrayed as massed formations advancing over hundreds of yards. Charles Saunders, an analyst with the Rand Corporation, provided a description of these French attacks:

> There are reports of attacks that were launched when the enemy was still five hundred meters away: Five hundred meters to advance over open ground, at the double if possible, carrying heavy packs and other gear, singing, screaming, eager to close with the enemy and apply the cold steel of the bayonet. The terrible result was predictable.[62]

This account implies that the advance was a bayonet charge over hundreds of yards, but historians have frequently failed to acknowledge that closing with the enemy and the bayonet assault are two distinct activities. The French emphasis on the offensive at all costs led to high casualties inflicted by new weapons technology in August and September 1914.[63] Historians have emphasized the close association between the bayonet and the pre-war concepts of "*élan*,"[64] the "outright offensive,"[65] and the "cult of the offensive."[66] Some historians have gone so far as to state the association plainly as the "cult of the bayonet."[67] Historians have also frequently concluded that the bayonet was to blame for these failures, proving that the bayonet was useless during the whole of the Great War.[68] However, French attack doctrine in 1914 was a far more complex issue than they have indicated. According to French doctrine, infantry attacks were to be supported by artillery and infantry fire to suppress enemy fire. This suggests that the high French casualties were the result of a failure of the supporting fire and training, rather than any inherent flaw in the bayonet as a weapon.[69]

This historical relationship between the bayonet and the long distance charge has carried over into British and, by extension, Canadian historical writing. For instance, Bill Rawling – citing the "lions and donkeys" historian Denis Winter – described the bayonet's role in the attack as "the crucial orgasm of the 200 yard charge."[70] Desmond Morton increased the distance in *When Your Number's Up* (1993):

> British Army tacticians believed that rushing the enemy with "cold steel" was the best way of moving soldiers across the last 300 to 400 yards of fire swept ground. The alternative – moving from one fire position to another – took too long and, they feared, gave a frightened soldier too many opportunities to duck out of the battle. Bayonet fighting might seem silly in retrospect, but it had a logical, if sacrificial purpose.[71]

These assertions that British Army doctrine relied on bayonet charges over hundreds of yards lack citations to primary sources; once again, the futility of the bayonet has simply been assumed to be true.

However, these critics of the bayonet have failed to distinguish the act of crossing No Man's Land from the bayonet charge. An examination of British infantry tactics in training manuals from 1902 to 1914 demonstrates a different conception of the bayonet charge than what has been advanced by either Rawling or Morton.[72] *Infantry Training 1914* emphasized closing on an enemy position by "bounding" in rushes toward their objective:

> The object of fire in the attack, whether of artillery, machine guns, or infantry, is to bring such a superiority of fire to bear on the enemy as to make the advance to close quarters possible... Infantry bounding forward until it [the enemy position] can be assaulted with the bayonet.[73]

Infantry Training 1914 broke the infantry attack down into three principles. First, *cover* had to be provided to permit the infantry to *close* on the enemy position. During the Great War, the term "cover" referred to using firepower to neutralize or suppress enemy fire, although it could also refer to environmental factors, such as terrain or darkness. After fire or cover had neutralized the enemy, the infantry then *closed* on the enemy position in bounds or rushes, using available cover or going prone. Finally, at roughly 30 yards, the infantry *charged* the enemy position.

These pre-war principles of the infantry attack were maintained throughout the war. The British *Memorandum on trench to trench attacks* of 31 October, 1916, emphasized the continued importance of the bayonet charge in the attack:

> The two fundamental facts which govern the modern assault are these viz: –
> a) The assault no longer depends upon rifle fire supported by artillery fire, but upon the artillery solely with very slight support from selected snipers and company sharpshooters.
> b) The decisive factor in every attack is the Bayonet.[74]

Even in light of the lessons of the fighting on the Somme, this document still placed the bayonet front and center in a system of combined arms that had formed the basis of British attack doctrine since 1902. The difference between pre-war and 1916 attack doctrine was the emphasis on the use of artillery to support the infantry advance by suppressing the enemy. This continuity between pre-war and wartime principles of the attack was emphasized in another memo sent to the 1st Brigade in May 1917: "It must be clearly understood that the pre-war manuals remain in force and that the instructions issued by G.H.Q. are merely amplifications of these manuals in order to meet the varying requirements of this campaign."[75] The principles of the infantry attack – cover-close-charge – and the manual from which they were drawn, *Infantry Training 1914*, remained the basic foundation of the Anglo-Canadian attack throughout the war.

The distance at which infantrymen were conduct the bayonet charge was also clearly stated in prewar training literature: in *Infantry Training 1911* as "about 30 paces"[76] which was roughly equal to the distance laid out in *Infantry Training 1914* as "about 30 yards."[77] This prewar range for the bayonet charge was reiterated in a Canadian training memo from December 1916 entitled *Bayonet Fighting for Platoon Commanders*, which maintained that "the fight is decided at THIRTY yards, just before the charge."[78] The bayonet charge, even before the war, was supposed to be a short final rush at the enemy. The long-range charges that did occur, under specific conditions, were not in accordance with proper training or tactics.

When the French attacks in the opening weeks of the war are examined within context of the British attack principles, the reason for failure becomes clear. With inadequate fire from artillery and infantry to neutralize enemy fire, the tightly packed formations attempted to close with the enemy positions only to be decimated before the bayonet charge could be launched. The opening attack of the Somme campaign--which has also been used to condemn both the use of the bayonet and tight formations--was a divergence from British doctrine brought about by a combination of poor training and an incorrect expectation of the effectiveness of the artillery bombardment. Commanders and staffs assumed that the smashing, week-long artillery bombardment would be sufficient to neutralize enemy machine guns and artillery, thereby making it safe for the poorly-trained attacking infantry to cross No Man's Land in easily controlled formations and occupy the destroyed German positions. The artillery bombardment failed to achieve its neutralizing task. The resulting military disaster had little to do with the bayonet.

VI. An Alternative Interpretation

Historians have paid little attention to the bayonet and interpreted the evidence to support what has become, through repetition, an easily quoted mantra. When investigated in detail, however, the arguments made to support the claim of the bayonet's obsolescence on the Western Front are far from conclusive. The use of wounding statistics as an indicator of bayonet usage addresses only a portion of the bayonet's effect. Alternative uses of the bayonet proved that soldiers were resourceful, rather than indicating anything about the bayonet's utility as a weapon. The acquisition of alternative weapons by soldiers contradicts the argument that close combat was a rare occurrence in the Great War--rather, it suggests that hand-to-hand fighting was a frequent occurrence. Finally, the image of the bayonet charge of 200 yards or more has failed to grasp the nuance of the bayonet's place as the final phase of the attack. What remains is the frequent repetition of the unchallenged notion of the bayonet's obsolescence.

This re-interpretation of the bayonet as a significant weapon system does not contradict the recent "learning curve" narrative. Many of these refinements affected the means to suppress the enemy in order to permit the infantry to advance to a position from which the bayonet charge could be launched. The challenge of the Western Front was in searching for the proper emphasis on each of the three principles of the infantry attack – cover, close, and charge – and the technology to deliver them.[79]

Chapter II: 1870 to 1914, The Bayonet Before the War

THE OFFENSIVE CHARACTER of the strategy and tactics of 1914 rested upon the concept of *élan*, which embodied both the ideas of fervent zeal and momentum. In strategic terms, *élan* emphasized the "positive aim"[1] of the attack, by assuming the offensive armies hoped to seize the initiative in the war. This idea greatly influenced the opening operations of the Great War. On the tactical level, the fervent zeal or "offensive spirit" was to drive soldiers forward during the attack with sufficient momentum and speed to minimize the duration of their exposure to defensive fire. Given the priority on closing with the enemy, the bayonet was an important weapon in the tactical application of *élan*. However, *élan*, and its emphasis on the bayonet, was a response to the increasing firepower of weapons in the late nineteenth and early twentieth centuries. In spite of *élan*'s emphasis on denser formations, more characteristic of tactics in the eighteenth and early nineteenth centuries, the motivation behind the use of these formations had changed considerably.

Élan was not the only response of military thinkers to the challenge of new technology. The armies of Britain, France, and Germany were also influenced by *dispersion* theories throughout this period; however, the challenges inherent in the application of *dispersion* meant that *élan* became more predominant. However, these two concepts were not diametrically opposed paradigms as many historians have suggested.² Like the dynamic changes occurring in infantry tactics, bayonet fighting also changed significantly between 1849 and 1914, and was not a weapon governed by mental stasis, as some historians have argued. As the doctrinal debates in the decades preceding the Great War demonstrate, military institutions were far from unresponsive and unchanging.

I. European infantry tactics 1870-1914: Paradigm or balance?

The Franco-Prussian War of 1870-71 demonstrated the significance of new technology on the battlefield.³ The Germans had won a relatively quick victory over the French Army, but the effects of new weapon technologies against large armies of reservists and conscripts proved that the role of firepower on the battlefield needed to be reconsidered. For example, at St. Privat the superior range of the French Chassepot rifle had torn apart tightly packed formations of Prussian soldiers before they were in range of their breech loading Dreyse needle guns.⁴ At Sedan the Prussians in turn decimated the French massed formations with their superior, rifled, breech-loading artillery.⁵ The result of these battles was a growing interest in theoretical and doctrinal writing as European military thinkers debated the nature of the modern battlefield and the increased firepower of weapons of the era. The basic problem was how to move the infantry across the final 400 yards of fire-swept ground in order to deliver the infantry assault with the bayonet.⁶ European theorists developed two solutions to the problem: *dispersion* and *élan*. However, these two ideas were not diametrically opposed paradigms as many historians have portrayed, in reality they coexisted in doctrine.

The first solution was to disperse infantry formations so they did not provide densely packed targets. In essence, *dispersion* had infantry act as skirmishing formations had throughout the nineteenth century, and as militia and irregular units with little drill training had for centuries before.⁷ Individuals and small groups of soldiers bounded using available cover and then developed a superiority of fire over the enemy as the number of soldiers in the firing line grew. The British use of bounding with fire and movement tactics is an example of *dispersion*. However, the theory of *dispersion* was not without its drawbacks. The most significant weakness of *dispersion* was the inability to gather the mass of troops necessary to achieve the quick and decisive results with the bayonet assault. Critics argued that the lower density of soldiers and the corresponding increase in the size of the battle space meant that battles would be longer and more drawn-out affairs with little ability for higher elements of command to influence the battle once it had been joined. Another tactical problem, observed by critics like Meckel and Maud'huy, was that soldiers could not be trusted to continue pressing forward after they had gone to ground – it was assumed that men operating in small groups lacked the morale to continue toward the enemy.[8] However, it was the practical application of *dispersion* that proved the most difficult hurdle to overcome. *Dispersion* required devolution of the control of battles to subalterns and NCOs possessing confidence, independence, and initiative. This required substantial faith in and training of lower ranks, which was a growing problem for armies in the decades before the war.

M. A. Ramsay observed a steady decline in the numbers of officers in the British Army in the early twentieth century. Ramsay concluded that the increasing number of vacancies for officers in the British Army made the devolution of control required for *dispersion* tactics impractical, due to the inability of the contracting officer corps to train and lead growing armies of short-term reservists.[9] Nor was this a strictly British phenomenon. Joel Setzen noted the same trend of a dwindling officer corps in the French Army, citing a 50% drop in the number of officer candidates turned out from the French military academies between 1900 and 1911.[10] David G. Herrmann also documented the acute officer shortage faced by the German Army in 1912 and 1913.[11] This left the implementation of *dispersion* tactics problematic from an institutional standpoint. Nor did the will exist to implement the radical institutional changes to facilitate *dispersion*; the acquisition of new technologies and the expansion of armies were proving themselves to be sufficiently expensive propositions.[12]

The other solution to crossing the final fire swept stretch of ground was *élan*, the use of offensive mass, spirit, speed, and morale to cross the final stretch of fire-swept ground to deliver the decisive bayonet assault.[13] At its heart *élan* was focusing on the imperative to close with the enemy this need to close meant that the bayonet assault and *élan* would be closely associated. Some of the early theorists discussing *élan* relied on the benefits of tight formations to morale and control, men were to advance on the enemy regardless of cover. However, many historians have come to associate *élan* automatically and erroneously with drill square formations advancing on the enemy. This, as shall be demonstrated, was clearly not the case.

The imperative to advance meant losses would be high in the face of the increasing firepower of modern weapons, but, through the willingness to accept losses, the infantry would have sufficient mass to deliver the final bayonet assault and achieve decisive and quick results. The primary appeal of *élan* was its expedience. Teaching formation drill and indoctrinating troops with the spirit of the offensive was a simpler proposition than the training in initiative and leadership required by *dispersion* tactics. These solutions suited the realities of government expenditure, a contracting officer corps, and the expansion of armies consisting of reservists and conscripts with increasingly shorter terms of service and less training.[14] Due to its expedience and the example of the Russo-Japanese War, *élan* and the "cult of the offensive" came to dominate European tactical thinking.[15] However, in spite of this, the infantry regulations of Britain, France, and German all advocated some form of fire and movement tactics for infantry in the attack.

However, these two solutions have been portrayed in historical writing as polar opposites. This is in part due to the nature of the community in which these theoretical works were produced for, which "valued forceful delivery, even to the point of overstatement."[16] This portrayal of *dispersion* and *élan* as polar opposites has also been caused by the hindsight of historians searching for correct and incorrect paradigms. For example, Francois Loyzeau de Grandmaisson, one of the 'Young Turks' of French military thought before the war stated to his students at the *Ecole Superieure du Guerre*:

> The French Army returning unto its traditions, no longer knows any law other than the offensive... All attacks must be pushed to the limit... charge the enemy with the bayonet in order to destroy him... this result can be obtained only at the price of bloody sacrifice. All other conceptions should be rejected as contrary to the very nature of war.[17]

This would seem to put Grandmaisson clearly in the *élan* camp. However, this is cast into doubt by his writings in 1906:

> In open ground a frontal attack by infantry under fire is impossible. In attack it is the task of the guns to constitute that fixed element of superiority of fire necessary to provide a barrier against the enemy so long as the labourious approach of the infantry lasts... it [the artillery] will proceed the every step of its infantry by its fire...[18]

Here, Grandmaisson was clearly advocating fire and movement, albeit a specialized form of it with the artillery providing the fire for the infantry to move forward in a manner echoing the instruction of the *Memorandum on trench to trench attacks* issued to the British Army a decade later. However, even the infantry advance he advocated is not a blind charge of the infantry with their bayonets at the point. Clearly the issue is not as black and white as it has been portrayed in the historiography.

The results of the fighting in the opening months of the First World War have, with the benefit of hindsight, permitted the vehement criticism of seemingly flawed tactics, and the terms associated with it – *élan* and the "cult of the offensive" – have become historically derisive ones.[19] The failures in August and September of 1914 have led some to conclude that *élan* and the bayonet was a useless weapon because of its close association with this cult of the offensive.[20] This condemnation marginalized the complex issues behind the emphasis on *élan* in European infantry tactics in 1914, as historians instead attempted to identify the correct paradigm of *dispersion* and condemn the incorrect paradigm of *élan* and massed formations.[21] Further complicating the identification of correct and incorrect tactical schemes is that these two sets of tactics were largely a question of balance between contrasting elements of doctrine: firepower versus shock, control versus devolution, and minimization of casualties versus rapidity of decision.[22] For instance, theorists identified as proponents of *élan* and shock like Balck, Du Picq, Grandmaisson, Foch, Joffre, and Haig advocated, albeit to a lesser degree, the use of fire to suppress the enemy in order to permit the assault.[23] Conversely, theorists identified with *dispersion*, such as William Nicholson, Petain, Boguslawski, Scherff, and Zeddeler,[24] still assumed the enemy position would eventually be assaulted with the bayonet. There was never a monopoly or complete neglect of either the bayonet assault or fire and movement.[25] The doctrinal and tactical debates between the 1870 and 1914 were, rather than a conflict between opposing paradigms, an attempt to strike a balance between the two opposing poles that best utilized the conditions and resources of military institutions.

II. 1871 to 1899: From Practice to Theory

Between 1871 and 1899, French, German, and British theories wavered between these two tactical poles.[26] The French emphasized *dispersion* tactics in the immediate aftermath of the Franco-Prussian War, and the 1875 French infantry regulations noted: "the impossibility of a body of troops of any considerable size to move or fight in close order, whether in line or in column, within the zone of the enemy's effective fire."[27] However, the French Army was also the scene of loud and aggressive doctrinal debate and French doctrine swung radically between the lessons of the Franco-Prussian war and the theories of Charles Ardant du Picq and his disciple Mikail Dragomirov.[28] For example, in contrast to the 1875 regulations, the 1884 infantry regulations, dispensed with the use of cover and instructed that the attack move forward with "head held high, unconcerned about casualties."[29] This was followed a decade later by regulations that emphasized the bayonet assault with soldiers in "shoulder to shoulder" formations, although this would not be the last major shift in French doctrine.[30]

As the victor in the Franco-Prussian war, the Germans had less impetus for immediate institutional change – their 1873 infantry regulations maintained tight formations in spite of the experience of the Franco-Prussian War. The German Army finally emphasized a radical form of *dispersion* in their 1888 infantry regulations. The German 1888 regulations went farther than those of other European armies and rejected the bayonet assault, preferring instead to envelope the enemy.[31]

The British Army saw a slower rate of growth than those of France and Germany and adopted a compromise between both *élan* and *dispersion*.[32] The British tactics increased the proportion of skirmishers while retaining tight formations in support to take advantage of the fire superiority gained by the skirmishing line to deliver the bayonet assault. By 1899, all three armies had been influenced by both tactical ideas. The first decade of the twentieth century offered two wars – the Boer War (1899-1902) and the Russo-Japanese War (1904-05) – that gave practical examples and further spurred the debate between *dispersion* and *élan*.

III. 1899 to 1914: From Theory to Practice

The open spaces of South Africa exposed a number of deficiencies in British tactics. During the conventional fighting phase of the Boer War (before March 1900), British attempts to assault Boer positions suffered severely at the hands of skilled marksmen armed with modern magazine-fed rifles.[33] The lethal Boer rifle fire forced the British to disperse infantry in the attack and utilize what meager cover could be found on the African veldt.[34] The tendency of the Boers to withdraw from threatened positions further frustrated British attempts to apply shock tactics.[35] As a result, the fighting typically featured numerically superior British forces attempting to envelope Boer positions and the Boers retiring to the next line of defense when sufficiently threatened.[36] The lessons of the war primarily emphasized the importance of *dispersion* to contemporary observers.[37]

The importance of the rifle and envelopment led to significant changes in the organization and training and a resulting devaluation of shock tactics in British military thinking, including substantial increases in musketry training for infantry.[38] This was most discernable in the cavalry, where for a short time the rifle became the primary weapon in training, replacing the traditional shock weapons of the sabre and lance. For example, the 1904 Canadian cavalry regulations went so far as to claim "for all practical purposes the profitable employment of cold steel is over."[39] The growing interest in mounted infantry during and after the Boer War also highlighted the increased emphasis on movement and firepower over shock.[40]

The findings of the Elgin Commission to assess British military performance in South Africa also observed the stresses developing between the particularistic regimental system and the expansion of the British armed forces. The commission concluded that the theory of the British system was sound, but over-commitment in South Africa and elsewhere in the Empire had led to serious flaws in practice. This led to the institution of "training" rather than "drill" manuals, the word training being used to distinguish between the previous regimental training schemes and the new uniform system under the direct control of the War Office. This standardized training of soldiers was to make up for the failings of the strained regimental system and the diminishing ranks of the officer corps.[41] The conclusions of the commission led the War Office to emphasize "individuality, initiative, morale, character, and the necessity for developing a resolutely offensive spirit."[42] The emphasis on independence and "offensive spirit" again expressed the British embracing both *dispersion* and *élan*, but the experience of South Africa had emphasized the need for greater *dispersion* and the ideas of devolution of control, "individuality" and "initiative."[43]

Germany and France also digested the lessons of the Boer War. In Germany, the lessons of the Boer War had reinforced the 1888 infantry regulations, which emphasized *dispersion* and envelopment to avoid the effects of modern firepower. The Boer War also inspired a short-lived interest in the study of "Boer tactics" in the German Army, which according to Bruce Gudmundsson had a lasting impact on a number of German officers who revived these tactics during the latter half of the Great War.[44] In France, the lessons of the Boer War led to another reverse. The 1904 regulations returned to a system of *dispersion* stressing the use of terrain for cover.[45]

The Russo-Japanese War of 1904-5 forced another series of re-evaluations, this time in favour of *élan*.[46] The bayonet assault featured prominently in Japanese attacks on the fortifications around Port Arthur.[47] European observers commented favourably on the superior morale of the attacking Japanese over the defensively minded Russians.[48] The Japanese successes gave the proponents of *élan* the practical example they needed to challenge the dispersion lessons of the Boer War.[49] However, according to Robert Engen, the lessons learned from the war were a case of selective observation. The Japanese had also combined *élan* and *dispersion*. They demonstrated their offensive spirit with their willingness to suffer high casualties in the assault; however, Japanese tactics were also heavily influenced by the *dispersion*-focused 1888 German infantry regulations. They advanced cautiously under the cover of night or siege works to bring soldiers as close to enemy positions as possible before launching the bayonet assault.⁵⁰

The selective lessons of the Russo-Japanese War and the expedient solutions of *élan* led to a diminished influence of *dispersion* in European tactics in years immediately preceding the Great War.⁵¹ For Germany, the Franco-Russian alliance of 1891 presented the problem of a war on two fronts. Their solution was the Schlieffen plan, which required knocking out France in a decisive, fast campaign, and then turning to face Russia. Japanese tactics of the Russo-Japanese War provided the model for the required decisive results, and the *dispersion* tactics of the 1888 regulations were replaced in 1906 by regulations that re-emphasized the bayonet assault as the means to achieve quick victories.⁵² In spite of this change, German regulations still had a clearly defined system of fire and movement for infantry in the attack.⁵³ However, this change did have its critics. Kaiser Wilhelm II had his own ideas on the merits of shock tactics: "The art of bayonet fighting is at present absolutely without value; no profit can be derived from it in war, and its study merely wastes time in peace."⁵⁴ This criticism questions both the historical portrayal of doctrinal debates of the late nineteenth century as a conflict between paradigms and the contemporary importance of such opinions expressed in military service journals and elsewhere. The German reversal of 1906 challenges the impact of personal opinions upon the actual determination of tactics and doctrine. If the titular head of the German Army could not make his will felt, it was unlikely a colonel in another army would meet with more success.

In France, the resistance of the *dispersion* of the 1904 regulations grew in strength as the Boer War faded from memory.⁵⁵ Ferdinand Foch, Noel du Castelnau, and Grandmaisson came to prominence in the French Army as the 'Young Turks' espousing the outright offensive. The 'Young Turks' dominated the General Staff and the military journals in France in the last few years before the war. However, as observed above, the historical perception that these theorists advocated simply hurling massed formations of infantry into the enemy fire is a gross over-simplification. Foch and Grandmaisson, for example, both advocated the combined use of fire and movement and *élan*. In any event, it is unclear what effect the opinions of the 'Young Turks' had on French infantry doctrine, the French Infantry Regulations of 1913 included the following instructions:

> Article 97 ...the current power of armaments makes impossible all daylight attacks in dense formation across open terrain. The offensive spirit can only be maintained by using formations which are as supple and as little vulnerable as possible. The infantry therefore fights as skirmishers...
>
> Article 109 ...Each of the fractions composing [the skirmish line] moves

forward by bounds [which] are as rapid and as long as possible... profiting by all those moments when enemy infantry fire is rendered less effective by the fire of neighbouring fractions or by that of friendly artillery...[56]

Clearly the outright offensive of the 'Young Turks' did not preclude the inclusion of *dispersion* in the 1913 regulations. *Dispersion* and *élan* were principles that co-existed in military doctrine.[57]

The vague language of British regulations has spurred historical debate on British doctrinal thinking in the years preceding the Great War. British Army regulations after 1902 continued to emphasize the *dispersion* concepts of bounding, fire and movement, as well as the "offensive spirit" required by the bayonet charge. In addition to the compromise between *élan* and *dispersion*, the language of British regulations refrained from defining specific lengths of bounds or the size of bodies of troops engaged in fire and movement tactics. Shelford Bidwell and Dominick Graham have portrayed this compromise as military planners simply avoiding the debate between *dispersion* and *élan*.[58] Tim Travers, and others, had taken this argument further and concluded that this lack of clear direction caused many British officers to adopt unofficially massed tactics and *élan*.[59] Albert Palazzo suggested that this ambiguity in regulations was a conscious decision on the part of planners to permit the application of a culturally-defined and flexible "ethos" in place of a dogmatic doctrine.[60] All of these arguments likely hold an element of truth. Ambiguity suited the varied environments, conditions, and conflicts brought about by the requirements of imperial defence.[61] The unofficial adoption of *élan* was surely influenced by the increasing likelihood of British Army involvement in a continental war in the years following the Russo-Japanese War. This British compromise demonstrated that *dispersion* and *élan* were not antithetical concepts; they coexisted in both prewar and wartime tactical doctrine.

Military thinking was far from static in the five decades leading up to the Great War.[62] The emphasis on "offensive spirit" in 1914 was not one based on the ignorance of military planners and theorists; it was a decision motivated by political and institutional expedience.[63] Nor was *élan* an idea that lacked merit. It is true, an overemphasis on *élan* led to horrendous casualties in 1914, but the historical condemnation of the invented "paradigm" of *élan* and its emphasis on the bayonet assault is incorrect. The effectiveness of both *élan* and the bayonet assault was amply demonstrated during the Great War.

IV. Training and techniques of British bayonet fighting 1849-1914

The factors influencing the debate between *dispersion* and *élan* and the tactical concepts themselves had an impact on the techniques and training of British bayonet fighting throughout the late-nineteenth and early-twentieth centuries. The simplification of bayonet fighting systems throughout the period reflected the influence of a growing Territorial (Reserve) Force after 1908 and progressively shorter terms of service for the regular army. The inclusion of bayonet fighting in the new "training" manuals after 1902 and the disappearance of private instructors and unofficial bayonet fighting manuals were in accordance with findings of the Elgin Commission, which identified the need for centralized control of training rather than training being the responsibility of the strained regimental system. As well, the increasing emphasis on *élan* in the British Army was demonstrated in the changes in bayonet fighting that occurred after 1905. Bayonet fighting, like infantry tactics, was not a static tradition when war came in 1914.

Charles Henry Angelo, the army superintendent for sword exercise from 1833 to 1851, developed the system of bayonet fighting used by the British Army in the late nineteenth century.[64] Angelo came from a prominent family of English fencing instructors that had established a respected school in London in the eighteenth century. Not surprisingly, Angelo's system of bayonet fighting drew on the language, analysis, and techniques of fencing.[65] Angelo's system involved the drilling of postures and parries, from which a variety of thrusts and points were delivered.*

Growing army interest in bayonet fighting prompted the publication of unofficial bayonet manuals. For example, celebrated writer and explorer Sir Richard Burton, who succeeded Angelo as the superintendent of sword exercise, published an unofficial bayonet manual in 1853. Burton expanded Angelo's system, and used a system of numbered thrusts, parries, and postures taken from the study of fencing.** In his work, he observed both

---

* Anglo's system of bayonet fighting was first published in 1849 as the official training manual and was expanded and reprinted in 1853 and again in 1857. Charles Henry Angelo, *Angelo's Bayonet Exercise* (Parker, Furnivall, and Parker, Military Library, 1853); War Office. *Angelo's bayonet exercise* (John W. Parker and Son, 1849); ——, *Angelo's bayonet exercise* (John W. Parker and Son, 1857).

** For those unfamiliar with fencing terminology a brief description of these terms may prove helpful. A "guard" or "posture" is a particular stance and body position from which a defensive action, a parry, or offensive action, a "thrust," "point," or "cut," is launched. The bayonet fighting in the second half of the nineteenth century classified the various guards and offensive and defensive movements according to their similarity to actions and postures of swordplay, which used a number to refer to each. For example, Burton's system of

the neglect of bayonet training in the British Army in the first half of the nineteenth century, as well as the stiff nature of Angelo's system, which was based strictly on drilling postures:

> The days have been when there was a prejudice against attempting to introduce into our armies a regular system of bayonet exercise. The feeling still lingers, however, amongst some officers of the different services, who oppose the innovation for a peculiar reason... The objections urged by them against bayonet practice are – that the men should be taught to depend solely upon the charge, when they have nothing to do but to keep together in line – that the real old English system is to thrust at the enemy without any other consideration but to run him through the body – and that the solider [sic] who is induced to rely upon his individual strength or skill would be more likely to leave the ranks, thus throwing them into disorder.[66]

Burton proposed a new system of bayonet training where soldiers, in addition to drilling postures, practiced these techniques against other soldiers. Burton proposed that "Loose practice should be encouraged, a wooden button covered with a leathern pad being fixed upon the point of the bayonet, and masks being worn to prevent accidents."[67] Further, he suggested the benefits of gymnastic exercise to bayonet training and organization of Corps level "Salle d'Armes."[68] Burton's system of bayonet exercise, while not codified in an official drill manual, proved influential. Bayonet fencing, which offered far more realism in training than merely practicing postures, became a staple of British Army bayonet training under the auspices of the Army Gymnastic Staff during the second half of the nineteenth century.[69]

Training in bayonet fighting in the second half of the nineteenth century was a synthesis of Angelo's official system of drilled postures, and Burton's system of "loose play." This involved pairs of soldiers in heavily padded armour and masks sparring with decommissioned or dummy rifles fixed with special spring bayonets – a light flexible blade similar to sport fencing weapons. As well, this was often done on a fencing piece, providing a thin line of attack and movement, emphasizing the holding of ground against an opponent while fighting in the narrow confines of a formation by restricting the movement of a fencer side to side, and to a lesser degree forward and backward. In general, the techniques of bayonet fencing at the close of the nineteenth century had changed very little since the requirements of pre-

---

bayonet fighting included three parries. These were labeled "prime," "tierce," and "carte," or first, third and fourth in English. There was not a "seconde" or second parry in Burton's system of bayonet fighting as the bayonet was incapable of an action corresponding to the "seconde" parry with the sword. Richard F. Burton, *A Complete System of Bayonet Exercise* (William Clowes and Sons, 1853), 28-30.

1850 close order tactics. However, bayonet fencing gave soldiers experience in sparring against an active adversary. The emphasis on drilled postures, unofficial manuals, and competitive sport was maintained throughout the remainder of the century. British Army drill manuals did not include any information on bayonet fighting, relying instead on occasional pamphlets and the efforts of private instructors like Captain Alfred Hutton.[70]

Hutton, a student of Angelo's fencing *salle*, continued the work of Angelo and Burton.[71] During the last four decades of the nineteenth century, Hutton was a privately hired fencing and bayonet instructor for a number of British Army units, and he published two unofficial manuals on bayonet fighting.[72] His system used the fencing terminology and the system of numbered guards, thrusts, and cuts established by Angelo and Burton.[73] However, Hutton observed the problems inherent in the use of elaborate jargon in teaching the infantryman bayonet fighting:

> In forming a plan of attack and defence for a military weapon such as this, it must be borne in mind to what class of persons the exercise is to be adapted. It is not intended for the amusement of the select few who frequent our London fencing-rooms, but for the general use of the infantry soldier, and must therefore be, as far as possible, devoid of technical terms; its cuts, guards, and points must be as few as possible, and it must be rendered so simple in its details as to be within the comprehension of the dullest recruit.[74]

Hutton's observation on the need to simplify bayonet fighting would be echoed twenty years later. The lessons of the Boer War and the findings of the Elgin Commission emphasized the need for uniform standards of training; this in turn led to greater centralized control over bayonet fighting. *Infantry Training 1902* included an appendix on bayonet fighting, unlike the British "Drill" manuals of the end of the nineteenth century that relied on separate, and largely unofficial, publications. The 1902 system of training further simplified the bayonet fighting system developed by Angelo, Burton, and Hutton and maintained the established pattern of drilled postures followed by "loose play."[75]

The Russo-Japanese war, and the practical example of *élan* it provided, forced a reappraisal of bayonet training during final phase of the preparation of *Infantry Training 1905*. The appendix on bayonet training was eliminated, and a refined system was published separately as *Instruction in Bayonet Fighting*.[76] The techniques of the 1905 bayonet fighting system were similar to those of the 1902 system, but the language was further simplified and more of the fencing jargon was dispensed with. For instance, in 1902, a soldier was trained in the use of the first parry (protecting the upper right side of the body), second parry (protecting the upper left side of the body), and third parry (a downward sweep protecting the lower half of the body)._77_ In 1905, soldiers were taught the same techniques, but they were now renamed the right, left, and "extra" (low line) parries._78_ The 1905 system also introduced an interim step to training between drills and "loose play," with the "wall bag" providing a target to practice delivering the point with the bayonet. However, this training device had to be improvised "by hanging a padded jacket on the wall about the height of the breast of a man on guard."_79_ *Infantry Training 1911* also suggested an alternative to the use of a padded jacket to make a 'wall bag,' suggesting a "suspended bag filled with shavings, etc., having a rope fastened horizontally behind it to prevent it swinging when hit."_80_ This "wall bag" became the iconic gallows sack on which soldiers practiced their bayonet fighting skills during the Great War.

The manual *Infantry Training 1905* also included a section on "The Practical Use of the Bayonet"_81_ which provided the following advice: "On getting to close quarters, select an opponent straight in front of you, and drive home a determined attack with the point of the bayonet, continuing the forward rush so as to close with him whether you are successful in bayoneting him or not."_82_ This demonstrated a departure from bayonet fighting techniques espoused by the nineteenth century systems. The emphasis on forward movement dispensed with the confines of the fencing piece so important to the nineteenth century conception of loose play and bayonet fencing. This also marked the association between the bayonet attack and the forward movement of attacking soldiers which was drawn from the lessons of the Russo-Japanese War.

The association of the bayonet with forward movement and *élan* motivated many of the changes in bayonet fighting in *Infantry Training 1911*. This new manual featured a shortened appendix on bayonet fighting that further streamlined the basic techniques. For example, the three points and a lunge used to attack in the 1905 system were simplified to a single attack of the *long point* in 1911._83_ *Infantry Training 1911* also elaborated on the association between the bayonet and forward movement, confirming a re-ascendancy of psychological "shock" effect, i.e. *élan*, and massed formations, by advocating:

> A bayonet charge will normally be delivered in lines, possibly many deep, against a defending force also in lines, over rough ground, which may be covered with obstacles. Single combat will therefore be the exception, while fighting in mass would be the rule. In a bayonet fight the impetus of a charging line gives it moral and physical advantages over a stationary line._84_

While this seemed to be a clear use of *élan* and massed formations, this charge was only to be delivered after the enemy had been suppressed by fire and the infantry had "advanced by rushes"[85] into a position where the charge could be launched.

Figure 4: "The long point" (Bayonet Fighting 1916).

Still, the success of a formation bayonet charge depended on the collective proficiency of individual actions. "Continuing the forward rush," as advocated in 1905, did create some practical problems for the soldier engaged in a bayonet fighting – a soldier who had missed his opponent and kept moving forward was to rely on his fellow soldiers to finish off the enemy now in close proximity.[86] *Infantry Training 1911* introduced three techniques for the soldier facing an enemy inside the range of the *long point*. During the Great War, these close techniques came to be known as *infighting*. The first was *shorten arms*: "Draw the rifle back horizontally to the full extent of the right arm, butt of the rifle either above or below the elbow."[87] This permitted the soldier to thrust the bayonet at a much closer opponent. The second technique of *infighting* was striking the opponent with the rifle butt, but that it "should only be used when the attack with the bayonet has failed."[88] The final technique was the use of tripping; however, the manual provided little detail, stating "Any easy form of tripping an opponent should be taught,"[89] and leaving the precise techniques up to the instructor. These techniques permitted soldiers to continue to attack their opponent after continuing the forward rush in accordance with *élan,* and techniques of *infighting* would develop rapidly after 1914 as soldiers gained experience in close quarters fighting.

Figure 5: "Shorten Arms" (Bayonet Fighting 1916).

V. Conclusion

Infantry tactics from 1870 to 1914 were not static; both European infantry tactics and British bayonet fighting reflected considerable institutional responsiveness to external influences. On the surface, infantry tactics still maintained the tightly-packed formations of the nineteenth century, but their purpose had altered significantly from maximizing firepower – which new weapons technology had significantly increased – to maximizing shock effect.[90] Bayonet fighting itself changed considerably in the years leading up to the War. The growth of *infighting* and the simplification of the British system of bayonet fighting and training after 1902 illustrated a growing emphasis on shock tactics in an army with an increasing proportion of short-term reservists.

Chapter III: Fear and Function

AS A WEAPON system, the bayonet presents few technical or mechanical features to explain its operation – as a result, it relies on the qualities of the soldier wielding it rather than any inherent function of the weapon itself. The soldier's perception of the relationship between distance and time is critical in understanding the dynamics and stresses of close combat. As distance between combatants reduces the pace of combat increases logarithmically, often to such a degree that soldiers often lose cognitive awareness in combat. This stress is fed by the instinct for self-preservation and the inherent awareness of the mortal danger resulting from close proximity of the enemy. Both the resistance to this stress and survival instinct must be challenged in order for the infantryman to engage in close combat, and, this is accomplished through training and conditioning. This resistance also forms the basis of shock effect, as opposing forces close on each other on the battle field the resistance grows and tests the resolve of soldiers on both sides. This creates of negotiation of sorts as each side tests its willingness to engage in close combat. Only if the confidence of both sides were roughly equal did hand-to-hand fighting occur; however, even if close combat did not take place, the bayonet still had an impact on the enemy in their decision to surrender or flee.

I. Close Combat: Instinct and Training

The relationship between time and distance has profound impact on combatants and is one of the core factors in the application of shock on the battle field. Understanding the relationship between time and distance draws upon fencing theory and a rudimentary understanding of the concepts and relationships between distance and time are critical in understanding the function of the bayonet. Bayonet training literature of the early twentieth century did not engage in theoretical discussions of fencing terminology; however, the nineteenth century systems of bayonet fighting on which they were based, and the creators of those systems, did. All three nineteenth century authors of British bayonet manuals were fencing instructors first and foremost and actively discussed the historical traditions and theories of close combat with the sword.[1] The first two, Henry Angelo and Sir Richard Burton, were also inspectors of British army sword exercise. The last of these authors, Captain Alfred Hutton, produced his works on bayonet fighting while actively researching and publishing works on the techniques of medieval and early modern martial arts.[2] Nor was Hutton alone in the use of these works in late nineteenth century military training. In 1898, Captain Cyril Matthey, an associate of Hutton's, published for the first time the second treatise written by sixteenth century English fencing theorist George Silver and argued its importance to the contemporary study of military sword exercise.[3] The centuries of analysis and traditions of swordplay formed the basis and origin of bayonet fighting in the early twentieth century.

Although theories of distance were not overtly discussed in bayonet training literature in the early twentieth century, specific ranges can be observed. Each corresponded to roughly the distance of a single step (roughly two feet). The widest measure was that of the *long point* (roughly six feet between combatants). Another step forward brought a bayonet fighter to the range of *shorten arms* or distance at which the rifle butt could be used (roughly four feet). A third step forward brought the combatant into range of "*the jab*," the use of the middle of the rifle, or the more distant techniques of *infighting* (roughly two feet). A final step brought the combatants into physical contact (*corps et corps*) where the majority of *infighting* occurred. Bayonet fighting itself took place at a variety of distances, each with its own techniques and its own corresponding impact on time in combat.

Time is the second concept to be understood. The most common understanding of time is length, or the amount of time taken to accomplish an action. In fencing parlance, this is referred to as *duration*.* However, this is not the only interpretation of the word "time." The translation of the French and Italian word "tempo" into the English word "time" fails to capture a number of additional nuances of the word when used in the technical language of fencing. The first is the potential of performing additional actions during the duration of another action, referred to as acting in *one time* or the *same time*. The simpler alternative of performing individual actions consecutively is referred to as acting in *double time* or *two times*. The second is of the pace at which offensive and defensive actions occur, referred to as *cadence* or *tempo*. The last of these technical meanings of the word "time" refers to the relationship between time and distance, in which as the distance between combatants reduces, the *duration* of actions also diminishes and the *cadence* of the fight increases.

To operate in the *same time* requires a great deal of training, discipline, and physical awareness. In general, as the British systems of bayonet fighting were simplified in the early twentieth century, the majority of techniques were performed in *two times*; however, some soldiers relying on instinct managed to act in the *same time*. One such instance is seen in Will Bird's bayonet kill on the evening of 16 November 1917, as the 42nd Battalion engaged in a minor operation near Passchendaele:

> It was like a bad dream. I hardly realized what I was doing. The officer had expected the Germans to surrender, and when one lunged at him with the bayonet he only escaped the thrust by falling to one side. Between his assailant and myself was the body of the *Feldwebel* killed by the pistol shot, and as I flourished my bayonet to bluff the German, he drove headlong at me.
>
> He tripped over his dead comrade. But I was too surprised by his ferocity to try and jump away. Instead I tried to ward off his weapon. Then I felt, not tearing steel in my own flesh, but the jolt of my bayonet as it brought up on something. The German groaned at once and sank down. I tugged the bayonet free and saw the fight was over.[4]

Bird managed to parry properly, leaving his bayonet in the correct line to be lunged upon by his opponent's attack bad footing. Bird had parried and, unintentionally, attacked in the *same time* as his opponent performed his single action in *two times*.

With training, a close combatant can intentionally manipulate the *cadence* and perform several actions within their opponent's *single time*. A

---

*Nicoletto Giganti, *The School or Theatre in which the diverse manner and ways of warding and wounding of the single Sword. And with Sword and dagger: are represented to you. Which must be studiously carried to exercise, & to be practically done in the profession of arms*. Translated by Aaron Miedema. (Manuscript, 2009), 11.

bayonet fighter could use a sudden increase in *cadence* to deliver a second action within the *duration* of his opponent's single action; this is referred to as *splitting the time, attacking in half time,* or *counter time.* Two techniques that *split the time* appeared in bayonet training literature during the war. The first, reintroduced to bayonet fighting in 1916, was the *feint,* where a soldier launched a false attack meant to draw a defensive response from the opponent operating in *two times,* then, increasing the *cadence,* the bayonet was disengaged under the opponent's parry and the attack with the point was delivered.[5] The second was the *duck,* appearing in 1917, where a false attack was followed by lowering the entire body to deliver the point underneath the opponent's weapon.[6] Although not overtly discussed in training literature, the inherent perception of the concepts of time in hand-to-hand fighting was of critical importance.

The correlation between time and distance is most significant to the role of bayonet fighting in the Great War; however, it is also the most abstract. As the distance between combatants diminishes, the *duration* of action also diminishes, or in other words the *cadence* at which actions occur increases as opponents move closer to each other.* In its most essential form, this relationship between time and distance is a simple application of the physical principle that it takes less time for an object to travel over a shorter distance.

Spurred by the instinct of self-preservation, the relationship between time and distance has a more profound and complex impact on the participants of a close combat.[7] The instinct of self-preservation and the corresponding tension and adrenaline also increases as the time taken for a threat to be delivered diminishes. This has the effect of further increasing *cadence* as an increasing amount of strength and speed are applied to actions in order to meet the increased pace of threats. Often, this increased defensive tempo can further increase *cadence* as combatants respond to decreased *duration* and attempt to deliver faster attacks in order to outpace

---

* Fencing theorist Ridolfo Capo Fero went so far as to use the word time to describe distance. Like the several distances or measures observed in bayonet fighting in the early twentieth century, Capo Fero also used several measures or distances in his analysis of close combat. Capo Fero's ranges were associated with rapier combat and corresponded to attacks delivered with the arm, the arm and body, and the lunge (involving the use of the arm, body, and foot), but his association of these distances with specific times has a direct application to bayonet fighting developed after 1905. Capo Fero applied a numerical value of time to each of these distances: the closest distance he described combatants as acting in half time, the second as acting in one time, the longest as acting in one and a half times. In his division, Capo Fero observed the relationship between time and distance in combative situations, albeit in an arbitrary manner. Ridolfo Capo Fero, *Gran Simulacro Dell'arte e dell'uso della scherma* (Sienna: Salvestro Marchetti and Camilo Turi, 1610), para. 49. Jared Kirby, *Italian Rapier Combat* (Mechanicsville: Green Bay Books, 2000).

the defense. Thus, as the distance between combatants reduces the *cadence* increases and so too does the stress level of the combatants, often logarithmically.

In the intense and close proximity of close combat, instinctual responses often take over from cognitive thinking. The responses of Canadian soldiers engaged in bayonet fighting hint at the blur induced by the stress of close combat. Will Bird, for instance, described his experience of bayonet fighting as: "a bad dream. I hardly realized what I was doing." This suggests that his actions in combat were detached from his cognitive control. Bird was by no means alone in experiencing this phenomenon. Thomas Dinesen, who was awarded the Victoria Cross for dispatching "12 of the enemy with bomb and bayonet,"[8] could only describe the events with the statement: "You live so intensely in the moment of the fight that you can hardly recollect any of the details afterwards."[9] However, complete disassociation was not the only response of soldiers to the stress of close combat. Jordan Crowe, with the 46th Battalion street fighting during the Battle of Amiens in August 1918, recalled with clarity one surprise encounter in which instinct took over from cognitive thinking:

> He [a German] stood there facing [me]... And I was carrying a revolver you know in my hand. And I never even thought of using it, funniest thing. I brought my knee up and hit him in the [blank] and he sank with a yell... Four others inside there were so frightened they disappeared... and it really threw me, funny. And I imagine if I shot him they would have started shooting back. So my stupidity could have also been to my advantage. The idea of knocking him out by hitting him in the [blank] was not new at all. But it was effective.[10]

It is not too much of a leap of logic to assume Crowe applied his knee to the groin, an *infighting* technique advocated in official bayonet training since 1916.[11] In spite of his ability to recount the experience, Crowe had little cognitive control over his responses at the time and failed to use the pistol in his hand. Adrenaline induced by surprise or desperation often forced soldiers to rely on instinctual and automatic responses. For Crowe, the automatic response seemed to come as a shock to the Germans as well.

An additional factor that served to increase the stress faced by soldiers engaged in hand-to-hand fighting was the increasing number of threats a soldier faced as they closed with the enemy on the Western Front. As a soldier advanced inside the range of the *long point* and *shorten arms*, he was forced to defend against the butt of the rifle and the middle of the rifle, at even closer distances the soldier faced additional possible attacks against which to defend: knives, sharpened entrenching tools, clubs, fists, feet, knees, head butts, and in some cases even teeth. As well, these threats are merely those of hand-to-hand fighting. In close combat, pistols could prove a devastating tool, and a soldier fortunate enough to have a round chambered in his rifle could dispatch an opponent quickly at bayonet range. New developments in military small arms, like the sub-machine gun and the bayonet equipped shotgun, demonstrated the importance of small arms in close combat, as well as emphasizing the significance of close combat in the Great War. Not only did the pace of threats increase as range diminished, but so too did the number of potential threats.

II. Close Combat Conditioning

Repeated practice of bayonet fighting techniques was only one part of training soldiers for close combat; mental conditioning of soldiers also played a significant factor in bayonet fighting on the Western Front. Bayonet training could to some degree overcome – or reprogram – self-preservation instincts in the crush of a bayonet fight. The instinct of self-preservation certainly had an impact on the way combatants fought, although it also presents the question of why combatants underwent the ordeal of close combat; the instinct of self-preservation equally informed a soldier to avoid such situations.[*] Training went some distance in establishing instinctual responses suppressing self-preservation, and even towards gaining some measure of control over the disassociation of cognitive thinking. The problem was that training, even intense training, was not entirely successful. That would have taken even more time than the army had already allocated to bayonet training. Will Bird, for example, had 60 hours of basic bayonet instruction in addition to regular practice on the assault course, but his cognitive thinking was still overwhelmed by the stress of the situation. In Grossman's example, in chapter two, the German lance corporal managed to maintain control over his cognitive thinking, although he still suffered from the effects of adrenaline – however, his training in saber dueling likely occurred over several years.[12] The alternative to years of combat training was the conditioning of soldiers for

---

[*] One such soldier was William T. New who claimed to have chosen to find a position in an ammunition column due to his fear of bayonet fighting: "I had a real abhorrence of both ends of the bayonet. I probably could have stuck one through those dummies they had on the drill grounds but I could not possibly have stuck one through another human being. So, if I were ever faced with the sharp end of a bayonet I would be in a terrible predicament. If I ran away, I would be charged with 'Cowardice in the face of the enemy' and shot. If I stayed and faced it, I would be an easy victim. So I decided I would not be any good in the Infantry, and that I would have to find something more suitable and where I would be more useful." William T. New. *The Forgotten War*. (Unpublished manuscript, 1982), 3.

close combat.

Conditioning also helped soldiers overcome the instinct of self-preservation in order to cross No Man's Land to come to bayonet range. More charitable critics have suggested mental conditioning for bayonet fighting prepared soldiers for the rigors of battle in general.[13] Less charitable critics have suggested this conditioning a fetish that led to atrocity.[14] When viewed from the perspective of examining whether the bayonet was an effective weapon, this fetish demonstrated the conditioning of soldiers to engage the enemy in the direst of situations. Dehumanization and emasculation of the enemy's prowess gave soldiers additional confidence in their ability to prevail and helped motivate them to close with the enemy. The confidence instilled by the combination of training and conditioning lay at the heart of the bayonet's effect as a weapon.

One of the most prevalent notions was the constant repetition of the idea that the enemy could not stand up to a bayonet charge by British soldiers; rather, they would surrender or flee. Another was associating bayonet training with a combination of hatred, dehumanization of the enemy, appeals to masculinity, and even sexual imagery.[15] These associations helped trivialize the task of bayonet fighting and motivated soldiers to kill the enemy in close quarters.*

Jingoistic propaganda helped to portray the German soldier as a mindless drone of abhorrent ideals, rather than a human being. Another aspect of dehumanization related the enemy to animals. Canadian infantryman Richard Rogerson recounted after bayoneting, bombing, and shooting fourteen of the enemy at Vimy Ridge that he thought "no more of murdering them than I used to think of shooting rabbits."[16] CEF soldier, Howard Charles Green, in response to the sinking of the *Lusitania* in 1915 used both emasculating and animal imagery:

> Those dirty sneaky [illegible]... of germans [sic] can't be content with fighting men but must fight women. If I could just feel myself putting a big bayonet right through one of the skunk's stomachs and rip him from toes to head. I'd be content to get one myself.[17]

---

* Of course these associations seemed unconvincing to some contemporary soldiers, an example is found in Gordon Reid. *Poor Bloody Murder: Personal Memoirs of the Frist World War*. (Oakville: Mosaic Press, 1980), 27 "We had some instructors from the British Army. One day while practicing bayonet fighting, jabbing dummy men, the instructor came up to me and said, "Put more into it! Just imagine these dummies are German and you hate Germans don't you?" I said "No, I don't hate anybody." He turned around to the men and said, "Blimey. Some of these Canadians are funny blokes!" One day later he came over and said, "I've been thinking about what you said the other day. I guess none of us really hates anybody!"

Dehumanization permitted soldiers to thrust their bayonets into animals or ideologies, not human beings.[18]

An example of these techniques appeared in *The Organization of the Bayonet Fighting and Physical Training in a Battalion C.E.F., 1916*:

> There is no braver man than the German soldier. He is ready at any moment to lay down his life for his country, but he is not the unit of the German Army. The N.C.O. is the unit of the German Army. Without the N.C.O. to lead and show the way the German soldier seems helpless. He lacks initiative and the fighting spirit, both have been suppressed. Where discipline and control play a strong part, at long range, he is an excellent soldier, but at hand-to-hand fighting when he has to rely on his own resources, he is lost.[19]

The next lecture continued on the same theme:

> The British soldier is at his best when he is fighting under no other control than that of his fighting spirit. This spirit is in his soul, it cannot be suppressed. During peace it finds relief in games of a fighting nature, boxing, football, etc. The German soldier on the other hand is suppressed; any spark of spirit is smothered by iron mechanical discipline. The fighting spirit of the German Army is supplied by the officer and N.C.O.
>
> It has been proved that when a German soldier sees a bayonet his fighting value is finished – he 'chucks up the sponge.' The Briton on the other hand when he 'sees red,' when his fighting spirit possesses him, gets more furious. He may quit the firing line with a slight wound when his blood is cool and he is hit at long range, but he is 'berserk' when his blood is up![20]

The German soldier was an inhuman automaton "smothered by iron mechanical discipline" and he would be beaten in close combat by the superior freethinking "Briton." Constant repetition of these images reprogrammed mental instincts, just as constant repetition of bayonet techniques on the assault course reprogrammed physical instincts.[21]

For a number of Canadian recruits, the combination of training and conditioning made them eager to meet the enemy at close quarters.[22] Thomas Dinesen, for example, recalled his eagerness when he first landed in France: "I shall see 'No Man's Land,' test the strength of my nerves under a bombardment and, I hope, get a chance of meeting the enemy face to face in a bayonet charge..."[23] Even after their baptism by fire, some soldiers maintained the eagerness wrought by the conditioning for close combat. Enos Grant, for instance, wrote home in 1915: "I am not afraid of them. The only thing I regret is that I couldn't get a chance to bayonet some of them, but hope to get the chance next time,* then maybe we will get some of our own back again."²⁴

One of the more curious aspects of close combat conditioning lay in the repetition of the concept of the "spirit of the bayonet." Canadian training literature frequently emphasized this spirit: "The spirit of the bayonet must be inculcated into all the ranks so that they go forward with that aggressive determination and confidence of superiority born of continual practice, without which a bayonet assault will not be effective."²⁵ Conditioning in the form of "aggressive determination" and training in the form "continual practice" bred the "confidence" for Canadian soldiers to push forward into close combat. This spirit may have also served to describe the blur of close combat when instincts took over from cognitive thinking. *S.S. 143 Instructions for the Training of Platoon for Offensive Action* openly acknowledged that bayonet training and fighting created an altered state in soldiers "bayonet fighting produces lust for blood."²⁶

Canadian soldiers observed this altered state. Thomas Dinesen described this trance as "yelling and roaring, with bombs and bayonets, battle-mad – regardless of everything in the world, our whole being intent on one thing alone: to force our way ahead and kill."²⁷ George V. Bell found a similar motivation "We are no longer human. Kill! Kill! KILL! That's our only instinct. With bayonet, bomb, trench knife, and even with our bare hands we kill."²⁸ In such situations, the soldier's bayonet and body

---

* Another instance of training-induced blood lust is found in the 5th Battalion billets on Salisbury Plain in the winter of 1914/5 appears in Baldwin, *Holding the Line*, 49-50:

"This man [a soldier named Bolous] was deeply interested in bayonet fighting, and would question our instructors until they loathed the sight of him. He studied the matter from all angles and would endeavor to get the man next to him to act the part of an attacking Hun in order to show us his own method of rendering Fritz *hors de combat*. Nobody ever volunteered as there is no knowing what he would have done in his eagerness to spit something with that bayonet. He devoured all that he could find in drill books about 'Hun Sticking.' He was particularly nerve trying at night, when we hobnobbed at cards or were reading before 'Lights out.' Everything would be quiet, except for the low murmur of conversation and an occasional heartfelt oath from a loser in the poker party. Then suddenly we would almost jump out of our skins, as a figure hurled itself at the rifle rack, seized a rifle from the stand, fixed the bayonet and rushed up and down the hut furiously parrying and lunging at an imaginary foe. Oblivious to everything except dispatching the figurative German, he would rush here and there while we endeavored to avoid the flickering steel. The man was enormously strong, and agile as a cat, and all we could do was to dodge as well we could until his paroxysm passed and he had settled down to work out some other scheme for Boche killing.

"We swore we would murder him if he did not cease these imitations of a madman, but glad are we all who knew him that we took his wild behaviour good naturedly, for a very short time afterwards he performed deeds of the most self-sacrificing kind under a wall of shell fire. Not a few men own their lives today to his devotion to duty on that awful day at Ypres."

seemed to take on a life and spirit of their own.

Conditioning was used in combination with bayonet training to give the infantryman sufficient confidence to enter into close combat. Training and constant practice permitted the soldier to reprogram their combative instincts to suit the bayonet and even overcome the disassociation of cognitive thinking; however, the frequent use of the rifle butt, observed by Grossman, suggested training was often insufficient to accomplish even the first task. The amount of training required for soldiers to ensure the retention of cognitive thinking during close combat was not within the means of the mass mobilizations of the Great War. Conditioning provided another way in which to encourage soldiers to advance into the stress of the bayonet fight. The constant repetition of the inadequacies of German soldiers in close combat and the prowess of the British bayonet instilled additional confidence. The dehumanization of German soldiers to the level of animals or ideological automatons helped to trivialize the task and motivated soldiers to exterminate an ideology.

III. The Negotiation of Close Combat

Defending soldiers had a variety of responses when presented with a bayonet charge, and to engage the enemy in hand-to-hand fighting was only one of them. The other options included flight, surrender, retiring to dugouts, or even playing dead. These options give credence to the popular notion that the German soldier was terrified of the bayonet, or more accurately the stress of close combat.[29] However, the decision of soldiers to fight, flee, or capitulate was not wholly dependent on the resolve of defending soldiers – attacking soldiers also had to demonstrate their willingness to engage in hand-to-hand fighting. The particular conditions of battle or doctrine could also have an impact on whether or not close combat took place. As a result, close combat was a negotiation of sorts between opponents and it demonstrated that the bayonet's significance went far beyond just running the enemy through.

The actual bayonet thrust was certainly tangible evidence of the bayonet's effect as a weapon, but it was not a case of Denis Winter's claim that "No man in the Great War was ever killed by a bayonet unless he had his hands up first."[30] Yet the killing of prisoners provides a starting place for the examination of the negotiation of close combat as it builds on the important issue of conditioning raised in the previous section. Given the blur caused by close combat and the blood lust inspired by conditioning, adrenaline, and rage, it was not surprising that attacking soldiers committed atrocities against surrendering enemies.[31] Conditioning likely led to the killing of prisoners; however, to suggest this was the only result of conditioning is erroneous – as it fails to consider how Germans came to be prisoners in the first place.

The notion that the bayonet instilled fear in the enemy hints at a closer association between the bayonet and surrender. Several instances within the records of the Canadian Corps suggest that the association between surrender and the bayonet was not a myth. There are numerous cases of Germans surrendering during hand-to-hand fighting.[32] There are also several instances during which the mere threat of close combat was sufficient to induce German troops to surrender – for example, at Hill 70, Corporal Dan Perman of the 10th Battalion:

> ...saw an enemy machine gun being put into action. He charged the gun and was shot through the right arm, but before the gun could be got into action he seized the barrel and tipped it over backwards down the sap from which it had been brought. An Officer and 6 O[ther]. R[anks]. then surrendered to him.[33]

Even though Perman did not attack the enemy directly, the threat of what followed, in spite of Perman being wounded and outnumbered seven to one, caused these Germans not to challenge their instincts of self-preservation. Here a lone Canadian soldier was close enough to attack the machine gun itself, and the word "charged" implied this proximity was due to the proper deployment of the bayonet in the attack. This is one of many instances found in the records of the CEF in which the use of the bayonet – or at least bayonet doctrine – was implied and not referred to directly in the capture of prisoners. For example, after encountering machine gun fire from a strong point on the Somme, Corporal Morrison with the 42nd Battalion "led the thirteen remaining men of the platoon on this position, stormed it with a rush and captured twenty prisoners."[34] How is this to be interpreted? The use of the bayonet and hand-to-hand fighting are not directly referred to. Even if hand-to-hand fighting did not occur, the use of the bayonet charge was implied by the use of the word "stormed" and "rush," since training and tactical literature maintained that the bayonet was a key weapon in the final push on enemy positions throughout the war. Regardless, the threat of close combat induced twenty Germans to surrender. This suggests the frequently repeated claim that the bayonet induced Germans to surrender was not a myth.

Tim Cook's article "The Politics of Surrender" examined the techniques used by German soldiers in the negotiation of surrender to members of the Canadian Corps, providing greater insight into the relationship between the bayonet and the act of capitulation. These techniques were acts of communication between combatants, and the negotiation required a close proximity. Surrendering soldiers pleaded for mercy, showed photographs of their families, or scrambled to help wounded Canadians, all within the range of a bayonet charge.[35] These acts of communication suggest the bayonet was involved in the negotiation of surrender and even induced capitulation. "Negotiated" surrender took place when bayonet attacks were imminent or had already been launched. An example of a failed surrender negotiation by German machine gunners at Passchendaele provides an example of the range at which the negotiations began:

> In all cases German gunners continued firing until the Toronto men got to within about 20 yards of them, then threw up their hands and attempted to surrender. However, the attackers, infuriated by the casualties these guns had inflicted on their comrades, gave no quarter but put the machine gunners to the bayonet.[36]

German machine gunners that kept firing until the final rush rarely found mercy at the hands of Canadian troops,* but the moment at which they attempted to surrender is significant.[37] The German's attempt to surrender when attacking troops were "about 20 yards" away corresponds closely to the distance at which a bayonet charge was launched expressed and reinforced by the training literature. Training literature from the CEF suggested the fight was decided by whether the attacking troops could organize a bayonet charge. *Bayonet Fighting for Platoon Commanders*, for example, indicated the point of decision in battle came when a sufficient weight of Canadian soldiers had advanced to within "THIRTY yards, just before the charge."[38] In the tense moments as attacking troops steeled themselves to rush the final yards, the defenders had the chance to contemplate the option of surrender and the possibility of summary execution or the grim hand-to-hand struggle that followed if they continued to resist.

There were other options available to defending troops faced with a bayonet charge. Many Germans retired to dugouts to await a counter-attack to wrest the position from the Canadians, or at least to give a chance for the adrenaline of attacking troops to subside before negotiating surrender.[39]

---

* Although sometimes continuing to fight could result in surviving the bayonet assault, Mr. H. R. Camp with the 18th Battalion recalled one such instance for the CBC's interviews for *In Flanders Fields*: "some of these German soldiers, they had guts too you know. I know one spot there, at Ballalemies we ran into a machine gun match, and we kept picking them off, one by one, and finally this fellow must of run out of ammunition. He stood up in his hole and started taking his gun to pieces and he was throwing the pieces at us, anything he could get a hold of. We knew then of course that he was out of ammunition and we up and rushed him. The officer bellowed, don't stick him boys...." LAC, RG 41, volume 10, 18th Battalion, Mr R. H. Camp, 2/7.

Some soldiers without a dugout to hide in resorted to playing dead to buy time for attacking troops to calm down or provide an opportunity for escape.[40] One such surprise occurred to a group of new drafts in the summer of 1917:

> One Heinie lay huddled in a corner and as I arrived I heard one of the new men in D company saying: 'I'm going to try my Bayonet in that chap. It can't hurt a dead man to stick him and I want to know what it feels like.'
> He posed his steel, ready to make the thrust in spite of several protesting voices, when the German yelled and jumped to his feet.[41]

Often the threat of close combat caused German troops to withdraw from their position – flight became particularly noticeable in the fighting during the last half of 1918. For example, a 2nd Battalion report from 23 August 1918 commented that "the enemy in not one case waited for assault but invariably retired before the attacking troops came to close quarters..."[42] Although a prevalent occurrence in 1918, Germans retreating in the face of the final rush occurred frequently throughout the war, therefore the breaking of German unit cohesion was a significant aspect of the bayonet's effect.[43] When presented with the option of facing the stress of close combat, many defending troops took to their heels out of self-preservation. However, often the Germans were not intimidated into breaking and hand-to-hand fighting ensued.

Confronted with the threat of close combat, defending soldiers faced a decision of fight, flight, or surrender. The soldier's confidence in surviving the encounter played a significant part in the decision making process. Training and conditioning gave soldiers the confidence to overcome the instinct of self-preservation and resist the bayonet charge. To surrender was the most direct option for soldiers who lacked confidence in their ability to prevail; however, the threat of summary execution (for which the Canadians had a substantial reputation amongst both German and allied formations) could motivate soldiers to face the alternative of braving artillery and small arms fire of attacking troops by retiring to the next line of defence.[44] Tactics and doctrine could also complicate the decisions made by defending troops. German defensive tactics, developed after the Somme, relied on localized control of defenses and counter-attacks as German soldiers fought from improvised strong points, withdrawing as particular circumstances of the Entente artillery and advance dictated.[45]

Entente attack doctrine also had an impact on the negotiation of close combat. From 1916 onward, German troops often faced long preparatory bombardments before being presented with the bayonet charge. These bombardments deprived defending soldiers of sleep, supply, and relief. This had a direct impact on the will to resist.[46] Such was the case of the German defenders facing the Canadian assault on Mount Sorrel in the early hours of 13 June 1916. The Germans occupied poor fortifications, shattered by previous fighting, and had been subjected to four days of bombardment before the bayonet charge was presented:

> The enemy opened up a short burst of rapid fire from their front trench, but it was ineffective; they then threw their bombs and the 16th men retaliated with bombs, for these weapons survived the mud; and then went in with the bayonet, at the sight of which the resistance collapsed. Many of the Germans appeared dazed. Most of them had no rifle and had no equipment on, and those who had, made no pretence of further fight once they saw the steel.[47]

Here, the exchange of hand grenades indicated the bayonet charge was launched properly, at 20 to 30 yards. The Canadians demonstrated their willingness to engage in close combat by launching the bayonet charge, and in response the dazed Germans surrendered. In the set-piece battle, artillery battered enemy positions and broke morale, but the bayonet charge and the stress of close combat forced defending troops to decide their own fate.

In essence, the negotiation of close combat was an act of intimidation, as attacking troops attempted to demonstrate a greater willingness to engage in the stress of close combat than the defenders. Will Bird described the process as a "bluff" when his platoon stumbled upon a German strong point during the fighting at Amiens in August 1918:

> We were in the open without a chance for cover. There was but one thing to do, bluff it. 'Come on,' I yelled, and put my bayonet level in front of me. We charged in a manner that would have tickled the 'canaries'* back in the Bull Ring at Le Havre – and not one German moved.
>
> Not until we were a few yards from them. Then they swarmed out with their hands high.[48]

These were two techniques used during the war to aid in the "bluff." The first is hinted at by Bird's account as he "yelled" to begin the bayonet charge, with the yell serving to communicate to Bird's comrades the course of action, but also alerting the Germans that the bayonet charge was being instigated. This relationship between the yell and the intimidation of the

---

* The term "canaries" referred to the yellow tabs that denoted training officers in British and Dominion armies. "Le Havre" was the location of one of the training facilities established in 1917 to test the training of soldiers upon their arrival in France, in a facility that was named the "bull ring." Before being sent to the front line a soldier had to demonstrate their competence in several elements of training, including bayonet fighting. Details of the course of study are found in: LAC RG 9 III, vol. 767.

enemy was acknowledged by *Bayonet Fighting for Platoon Commanders*: "You have then already 'put the wind up' the Hun, and HE WILL GO when you YELL 'CHARGE' and charge."[49] This use of a yell at the beginning or during the bayonet charge had been advocated in official bayonet training literature since 1915 and remains a central foundation of bayonet training today.[50] The yell, shout, or cheer also frequently appeared in the records of Canadian operations throughout the war.[51] The second technique was the "*killing face*," which was meant to convince the enemy of the intention of the soldier to kill in close quarters.[*] The killing face was espoused by Colonel Ronald B. Campbell's ubiquitous "bayonet circus." Campbell's circus toured the Western Front, beginning in the spring of 1916, giving lectures on the "spirit of the bayonet" and demonstrations of bayonet fighting techniques meant to train and condition soldiers.[†] The basis of the negotiation of close combat was one of bluff and intimidation, in which soldiers demonstrated a disregard for their self-preservation instinct.

This demonstration forced the other side to decide whether or not to engage in close combat. The German "fear" of the bayonet, appearing frequently in the literature and accounts of the Great War, was not a myth. The "fear" of the bayonet was an acknowledgement of the decision forced on any soldier when close combat was offered. To be fair, this was not because the Germans were inferior soldiers. The Germans were generally on the defensive on the Western Front. The bayonet, being an offensive weapon, received less emphasis in German operations as defensive tactics relied on firepower to prevent attacking troops from closing to bayonet range.[52] However, the German tactics of rapid counter attacks into positions overrun by the Entente assaults meant the bayonet and the negotiation of close combat maintained an established place in German doctrine.

The power of the bayonet lay not in just running the enemy soldier through – the bayonet charge forced the enemy to question their confidence and determine if close combat would occur. The growing firepower wielded by armies in the Great War may have inflicted heavy casualties and cracked

---

[*] The killing face did not appear in training literature, but began to appear informally in training in 1915. David McMillan. *Trench Tea and Sandbags*. (Canada: R. McAdam, 1996), 6; George C. Machum, *The Story of the 64th Battalion, C.E.F.: 1915-1916*, (Montreal: Industrial Shops for the Deaf, 1956), 58-60.

[†] Colonel Ronald Campbell was a British born immigrant to Canada, he served in the Boer War with the Royal Canadian Regiment before being commissioned in the Gordon Highlanders in 1900. He was appointed the Assistant Inspector of Bayonet Fighting and Physical Training in 1916 and operated a training facility in France and toured front line units giving bayonet fighting demonstration and lectures. John G. Gray, *Prophet in Plimsoles: An Account of the Life of Colonel Ronald B. Campbell* (Edinburgh: Edina Press, 1976), 21.

morale, but forcing defending troops to break or surrender fell on the bayonet charge. We know the statistics of soldiers wounded by the bayonet, but the number of soldiers killed by bayonets will not likely ever be known for certain. It is, however, safe to assume that accounting for the number of soldiers killed by the bayonet would raise the proportion of casualties inflicted by the weapon. For defenders who did not stand against the charge, the bayonet rightfully deserves credit for a substantial proportion of the more than 42,000 prisoners taken alive by the CEF and the unknown number killed during the process of capitulation.[53] The range at which surrender was negotiated – often face-to-face – implied the bayonet, or the threat of it, was involved in the act of capitulation. A fourth category must also be considered, even if it is unquantifiable: soldiers induced to flee in the face of close combat. Often those soldiers, once in the open, became victims of small arms fire and artillery. In these situations, the bayonet rightly deserves at least a portion of the credit in routing the enemy. The bayonet was a simple weapon whose primary function was in breaking the cohesion of defending troops, first through intimidation and, failing that, through the grim task of hand-to-hand fighting.

Chapter IV: 1915, The Bayonet and Trench Warfare

I. The Challenge of Trench Warfare

THE ASSASSINATION OF Archduke Ferdinand in Sarajevo on 28 June 1914 set off the wider European conflict that politicians and soldiers had been preparing for over the previous decade. According to prewar plans, the armies of Germany, France, and Britain hurled themselves at each other, each seeking to gain the advantage of the attack and decisive victory. On the Franco-German border the French attacked in accordance with Plan XVII only to be repelled with grievous losses in the Battle of the Frontiers. The Schlieffen Plan proved more successful, with the Germans advancing through Belgium into France. However, the tight schedule of the plan began to unravel as the Germans strained their supply lines and Belgian resistance proved more stubborn than anticipated.[1] The French armies recovered and, with the assistance of the British, stopped the German advance at the Battle of the Marne. With their offensive momentum stalled, the Germans sought to turn the Entente flank by advancing along the English Channel. This "race for the sea" ended with the successful, but costly, resistance of the British Army at the 1st Battle of Ypres in October and November. By the end of November, the offensive energy of the armies had been ground away by the tremendous casualties incurred in the fighting. The adversaries then settled into defensive positions along a line running from Switzerland to the channel.

For the sake of survival, the infantry entrenched; this act alone likely ensured the utility of the bayonet.[2] At first, these trenches were mere scrapes in the ground to protect soldiers from enemy fire, but these initial defenses developed into ditches, then trenches, and then into elaborate interconnecting networks. The use of large formations in the attack gave way to a rhythm of manning the fortifications in shifts, as the defense could be conducted by a smaller number of soldiers. Soldiers rotated out of the front line were now able to reorganize, retrain, and be used in the manpower-intensive activities of localized offensives or maintaining the massive network of trenches. The Germans, who adopted a defensive posture in France, took the opportunity to establish themselves on favorable ground and led the way in developing fortifications.[3] Entente armies, compelled to attack in order to drive the Germans from their territorial gains in Belgium and northern France, were slower to develop their defensive works. However, given the complexities inherent in offensive operations, they too were forced to adopt a static posture by the end of 1915. Belts of barbed wire also expanded in front of the trenches. The use of barbed wire was by no means new in military operations – it had been used in the Boer War and the Russo-Japanese war – but the scale of the siege lines saw an unprecedented use of barbed wire. Like the *abatis* in late-medieval siege works, barbed wire was intended to prevent the use of shock tactics by disordering and slowing the movement of attacking troops.[4] Use in conjunction with modern technology, it crowded attacking troops into narrow gaps in the wire that served as killing zones for well-sited machine guns.[5]

While the infantry had moved underground to avoid the worst effects of modern firepower, artillery was forced to move away from the fighting to avoid fire from modern rifles and machine guns, which could harass the enemy at up to 2000 yards. Artillery now had to fire at targets beyond their line of sight.[6] The two methods of firing at unseen targets were observed fire, in which a forward observer in communication with the artillery could view and correct to fall of shells, and predicted fire, in which the artillery fired shells by using only a map for reference. Both techniques had been practiced before the war, but between the complexities of communication and accurate map-making, the refinement of these techniques was a slow process of trial and error.[7] In spite of refinements in firing beyond the line of sight, the trench still proved a small target to hit directly and continued to offer defending soldiers substantial protection. However, attacking soldiers, forced to leave their entrenchments, suffered heavily at the hands of the artillery. Even if the attacking troops succeeded in securing a portion of the enemy line (the safest place for attacking troops after going over the top), the immediate support and supply of these new positions was still subject to artillery harassment, as they too had to leave the safety of entrenchments.

The trench provided substantial protection against low trajectory weapons and all armies were quick to adapt techniques and weapons suited to the conditions of siege warfare.[8] Before the close of 1914, armies began to develop and expand their arsenals of high trajectory weapons to penetrate defensive works. The Germans had the lead in these weapons as they began the war with a greater number of heavy howitzers and mortars.[9] In contrast, the French and British had placed an emphasis on low trajectory field guns before the war, neglecting these now-important weapons.[10] The British Army had deployed a handful of eighteenth century-pattern Cohorn mortars to Europe, but they began improvising mortars by the end of 1914.[11] By the fall of 1915, the industrially produced Stoke's mortar would begin replacing these improvised weapons. The hand grenade had also been deployed in small numbers with BEF sappers in the summer of 1914, however, with the onset of trench warfare the demand for these weapons increased greatly.[12] Soldiers resorted to manufacturing their own grenades out of whatever resources they found at hand. These improvised weapons were acknowledged quickly by the chain of command and assimilated into doctrine. By March 1915, even before an industrially-produced weapon was available, specialist grenadiers and grenade tactics and training regimens had been established in the British Imperial Army.[13] Both weapons and tactics responded quickly to trench warfare. It was in to this rapidly changing battle space of the Western Front that the Canadian Contingent appeared in the spring of 1915.

II. From Canada to France

The British declaration of War on Germany included Canada, and on August 4th 1914 Canada was at war with Germany. During the second half of August, the Canadian Army increased to ten times its pre-war establishment as over 30,000 volunteers for the Canadian overseas force gathered at Valcartier camp to begin training. Nor did the pace slacken significantly for two years – the Canadian Army grew to over a hundred times its pre-war size by 1917. Such growth made uniform training in even basic skills difficult.[14] Training was further impeded by the deployment of much of Canada's experienced military personnel overseas – the veteran men of the Princess Patricia's Canadian Light Infantry (PPCLI) embarked for England at the end of August 1914. As well, the only regular formation of the Canadian Army, the Royal Canadian Regiment (RCR) embarked for Bermuda at the beginning of September.[15]

In spite of the apocryphal story of Sam Hughes giving impromptu lessons in bayonet fighting at Valcartier in the summer of 1914, the training of the First Canadian Contingent in bayonet fighting was the responsibility of the British Army.[16] Bayonet training did not begin until the Canadians arrived at Salisbury Plain in November, when they were given five Imperial bayonet instructors to conduct training.[17] Canadian soldiers beginning their bayonet training in November also corresponded to the syllabus of *Infantry Training 1914*, which did not introduce the soldiers to bayonet fighting until the seventeenth week of the twenty-four week training syllabus.[18]

The five Imperial instructors were given the task of delivering 20 hours of training in bayonet fighting to each of the 17,000 infantrymen of the First Contingent. To accomplish this, four NCOs were selected from each battalion (one per company) and were trained as assistant instructors. These assistant instructors were then responsible for training their respective companies. This two-tier system for training remained the basic foundation of bayonet training for the course of the war, although it underwent considerable refinement and expansion by 1918. The delegation of responsibility created a grassroots training system within each battalion. This had the advantage of permitting the limited number of instructors to train vast numbers of men; however, the system also produced highly variable results.*

The development of training institutions and qualified instructors proved problematic under the conditions of mass mobilization of 1914 and, in relation to the bayonet, continued to present a significant difficulty until the end of 1916. The sheer scale of the growth created tremendous strains on the training systems, as what few qualified instructors there were strove to train more instructors to deal with the sudden influx of manpower.[19] Bayonet training presented additional problems aside from the lack of instructors. The pre-war system of training using "loose play" and bayonet fencing could not cope with the growth; the amount of equipment required – masks, padded armour, and dummy rifles – could not be supplied to maintain the old system of training. "Loose play" was eliminated, and the

---

* This grassroots system of training was one of two training systems of training adopted by Anglo-Canadian forces during the war; the other system was the establishment of training select specialists. The specialist system of training was used primarily for new weapons and examples are found in the assimilation of Grenades and Lewis guns into the Canadian Corps. For weapons and equipment universally required by infantrymen the grassroots system was typically used, another example of such a system is seen in the training of Canadian infantry in gas warfare, see: Tim Cook, "No Place to Run" (M.A. Thesis, Royal Military College of Canada, 1997), 120-1; Cook, *No Place to Run*, 61. Ramsay, *Command and Cohesion,* 170.

final stage of training focused on the pre-war "wall bag," which now became the bayonet sack hung on a gallows or lain on the ground (for practice against entrenched soldiers). This new training became known as "assault training," with the ground over which the soldiers advanced called the "assault course." The precise evolution of the wall bag to the gallows sack and the assault course is undocumented in Canadian sources, but the sack had become firmly established by December 1914.[20] This meant bayonet training now consisted of only practicing postures and stabbing sacks; soldiers were no longer gaining experience in dealing with a target that was fighting back. Thus, until the introduction of "blobstick" in 1915 and 1916, soldiers had no official means for practicing the parry or defence in bayonet fighting.

Figure 6: Men of the 29th Battalion practice the art of Bayonet fighting with wooden pugil sticks

In February of 1915, the Canadian Division completed the remainder of its basic program of training – including bayonet fighting – and was moved to France to begin acclimatizing to the conditions of trench warfare, this included additional training in the new tactics of bombing and trench clearing.[21] Each brigade established a company of specialists in the manufacture and use of hand grenades; in addition, each infantry company also sent 16 men for specialist training in bombing.[22] Bombers were to be used in minor operations against enemy trenches, posts, saps, as well as for the close defense of positions. In major offensive operations bombers were organized into "Trench Storming Parties"[23] where the bomb and the bayonet became closely associated. In spring 1915 the *Memorandum of the Training and Employment of Grenadiers* laid out the basic tactics of bombing down trenches:

> On arriving at traverse 1, the bayonet men should place themselves in position AAA, the N.C.O. at C, or as required, the grenadiers at BB, behind the traverse with the carriers, if any, and spare bayonet men behind them. No. 1 Grenadier then throws a grenade over the traverse into trench X, and a second one in trench Y. The leading bayonet man can then move forward, so as to see into trench X. If it is clear he passes back word, and the 3 bayonet men move up trench X and occupy positions at traverse 2, similar to those at traverse 1. The grenadiers then follow, and throw grenades into Y and Z. Until Y is clear, the reserve bayonet men remain behind traverse 1, in case the enemy should throw grenades into trench X.[24]

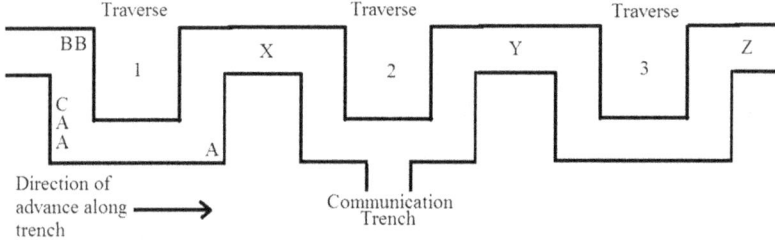

Plan for grenadier training

However, in battle trench storming parties did not always have the leisure to throw grenades until the next bay was clear, nor did the enemy obligingly remain in place to have grenades thrown at them. As bombing parties advanced the bayonet men leading them were called upon to engage the enemy at close quarters. In a letter home in May 1915, Second Lieutenant E. L. Yeo noted the importance of the bayonet in "bombing down" trenches:

> The system often used by us in capturing trenches is also interesting. Following a heavy bombardment of a portion of the enemy's line a bayonet charge is made on the same. A footing is thus made and a bombing party then comes into action. The bombing party consists of a number of men armed with hand bombs who are immediately preceded by others with fixed bayonets. Bombs are hurled over the heads of the latter people at the enemy, the demoralized survivors of which are summarily dealt with by the bayonet men.[25]

After an enemy trench system had been penetrated by the initial bayonet assault, small bombing parties led by bayonet men fanned out along the trenches.[26] Bombs were thrown into the adjacent sections or bays of a trench and then followed by bombing specialists trained as bayonet men who rushed the bay and finished resistance.[27] An unnamed soldier of the 44th Battalion described bombing down a trench in more routine terms in 1918 as "a ding dong affair – a few bombs into a bay or two, then rush in with bayonet; then repeat."[28] During the war, the bayonet saw considerable use in "bombing down" trenches – the main method for ousting the enemy from his defences.

III. Ypres: Baptism of Fire

The 2nd Battle of Ypres is remembered largely as a defensive battle for the Canadians, who in spite of high losses, chlorine gas, and faulty Ross rifles managed to hold back German attacks between 22 and 25 April, 1915. In spite of these significant characteristics of the battle, the bayonet still saw considerable use.[29] The collapse of the French Divisions on the Canadian left flank on the afternoon of 22 April forced two Canadian counter-attacks in order to stop the breach in the entente lines, with these attacks including the first Canadian bayonet charge of the war.

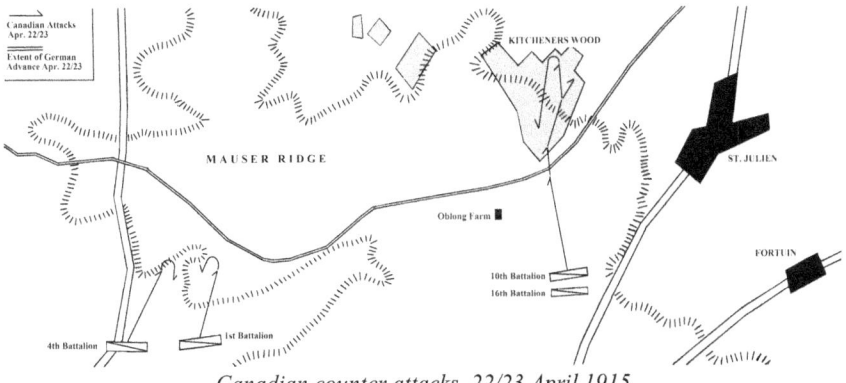

Canadian counter attacks, 22/23 April 1915

Late in the afternoon of 22 April, the German's unleashed an intense bombardment on the city of the Ypres. The bombardment was followed by the release of chlorine against the 45th Algerian Division on the Canadian left flank. Almost immediately, the Algerian line melted away under the gas attack, leaving a six-kilometer gap in the Entente lines. By 6:00 p.m., the Germans had driven three kilometers into the Algerian line toward Ypres and began entrenching. The Germans positions rested on Mauser Ridge overlooking the town of St. Julien, which flanked the 13th Battalion holding the front line and the rear areas of the Canadian 3rd Brigade.[30] The Canadians had to counter-attack quickly in order to relieve the threat to their rear areas before the Germans had the opportunity to consolidate the position. The objective of the first Canadian attack was Kitcheners Wood at the east end of Mauser Ridge. The 10th and 16th Battalions were ordered to counter-attack at 10:00 p.m. The stage had been set for the first Canadian infantry attack and bayonet charge of the war.

In the anxious minutes before the advance, officers of the 16th Battalion engaged in last minute drilling, which was to have a significant impact on the execution of the attack. Bernard Charles Lunn with the 16th Battalion recalled this drilling:

> Then we got the order to move, well we were over I would say a kind of ploughed field, a ploughed field, ordered to advance, fixed bayonets, practicing we thought fixed bayonets, lying down, standing up, on fixed bayonets, lying down, fix and unfix bayonets, well it began to be a joke. We wondered what the heck... I think, the officers they knew that there was evidently was something going to take place... we kept moving up about 100 yards at a time, flopping in the same old system fix, unfix bayonets, just to keep the boys interested...[31]

The battalion history supports Lunn's account, but it records the bounds as being much shorter moving 20 yards then going to ground.[32] The officers of the 16th Battalion were engaged in the last minute drilling of their unit in the proper technique of bounding toward their objective, as advocated by *dispersion* tactics and in British training manuals since 1902. The motives behind this last minute drilling are unclear; perhaps the intention was to preoccupy the soldiers in routine tasks in order to distract them from the tension of waiting for the attack, as was suggested by Lunn. Another motivation may have been the fear that the men, facing their first experience of battle, would not be able to properly apply their training, and that a crash course was required. Either way the officers of the 16th Battalion were attempting to apply the bounding techniques espoused by their training.

The attack went forward at 11:48 p.m., the Canadians beginning the 500-yard advance to Kitcheners Wood with the 10th Battalion leading. The 10th Battalion diary records: "Order to advance given, not a sound was audible down the long waving lines but the soft pat of feet and the knock of bayonet scabbards against thighs... a hedge was unexpectedly encountered and the noise of breaking through brought a hail of bullets, rifle and machine gun."[33] This hedge lay roughly 200 yards from the German positions and played a significant role in the success of the attack. However, according to Sid Cox of the 10th Battalion, the hedge line did not serve as a noisy impediment for the attack; rather it served to conceal the first 300 yards of noisy advance of the untried Canadians. Cox recalled: "Well, I don't think it [crossing the hedge line] caused any more noise than we were making. We were so all mixed up that nobody knew what a bayonet attack [was], [or] who anybody was, you could hear somebody, 'Hey Canada, Hey Canada, Come here.'"[34] Whether or not the hedge served as a noisy impediment to the advance or hid the Canadian attack, the Germans, now aware of the attack, sent up a flare and began firing with rifles from the woods and with machine guns on the Canadian left flank.

Now under fire, the Canadian battalions closed on the woods and demonstrated the impact of the last minute drilling by the 16th Battalion. Duguid's official history suggests both battalions surged forward in a mad rush over the last 200 yards.[35] However, this characterization of the attack may not be correct; Captain H. A. Duncan described the 10th Battalion charge in a letter home:

> When we were within three hundred yards of the trench we came upon a thick hedge, and after some delay we managed to get through. The fire was getting quite hot. From the hedge we made a rush of about fifty yards. By this time they had spotted us and the fire was awful, coming, as it seemed, from all directions, making a steady roar. We pushed forward another hundred yards or so, and when the fire slackened for a moment the front line charged, followed by the second line about twenty yards in the rear. We bayoneted the Germans who remained in the trench and chased the balance who made for the wood in the rear of the trench."[36]

Duncan's use of the word "rush" may suggest that the 10th Battalion used two bounds (first of fifty and then of 100 yards) before launching the bayonet charge. Duguid himself recorded that it took two minutes for the Canadian to cover the last 200 yards to Kitcheners Wood, which indicates the advance was more complicated than simply a flat run over the final stretch of ground. At fifty yards from the woods the 10th Battalion launched the bayonet charge and overran the German position, proceeding to execute "brisk work with bayonet and butt."[37]

For the 16th Battalion, the last minute training was effective and the use of bounding tactics was clearer. Charles Lunn recalled the advance of the 16th Battalion: "You'd get up, stoop up, and perhaps move 20 yards down again, the order of the officer, down, up again, on again, and it was from then on until we got within 50 yards, shall we say, 50 yards of the wood. [Then] everyman for himself, of course."[38] The 16th Battalion diary supports Lunn's account of events after the firing started: "We then doubled and when flare went up [we] lay down."[39] The 16th Battalion used shorter and more orderly bounds as they had practiced just before the attack began. The manner in which the 10th and the 16th Battalions advanced at Kitcheners Wood demonstrated the latitude offered infantry commanders by the loose principles of the "ethos" governing British infantry tactics, as well as the lack of uniformity in tactics that could arise.

In spite of a lack of support, hasty planning, and high casualties the Canadians had attacked and taken their first position from the enemy at the point of a bayonet. The cover of darkness and the hedge line gave the Canadians the ability to close to a distance where they could use speed to overwhelm the hasty German defenses; however, this still proved costly and of roughly 1,600 men who began the attack, only 458 remained in action the next morning.[40]

Kitcheners Wood contrasted sharply with the counter-attacks carried out by the 1st and 4th Battalions, six hours later, against German positions atop the west end of Mauser Ridge. Here an attack was conducted in daylight over 1,500 yards of open ground. Both battalions employed proper bounding techniques, but this could not make up for the fact that the soldiers had to advance across open ground in full view of the enemy and with insufficient suppressing fire.[41] Even the line of trees 600 yards from the German positions could not prevent the predictable result, and the Canadian attack was forced to ground, 400 yards from the German line, with heavy casualties.[42] The Canadians never made it to bayonet-charging range.

IV. Festubert: The Problem of the Attack

Both Kitcheners Wood and Mauser Ridge were hasty counter-attacks and were not indicative of the set-piece operations in which the Canadian Corps would later excel. The first Canadian set-piece attacks occurred at Festubert and Givenchy in May and June 1915.[43] While the bayonet was a key element of the Canadian set-piece attack, a successful assault first had to reach the enemy, otherwise the bayonet charge could not take place – as illustrated by these two battles, several factors could prevent the main assault from reaching the enemy trenches.

The first Canadian attack at Festubert occurred on 18 May; the 14th and 16th Battalions were ordered to support a larger operation by the British 7th Division. As these battalions had only arrived in the line the day before the attack, they had little time to reconnoiter the ground over which they were to advance – navigation through the shattered environment of No Man's Land was to prove a significant factor in the forthcoming battle. The

1915, The Bayonet and Trench Warfare 69

objective was the German front line trench roughly 900 yards east of the British line. In the path of the advance was an orchard 200 yards short of the final objective. Poor quality maps made the identification of objectives and artillery targets problematic, leading to the two hour bombardment meant to suppress the German positions failing to find its targets and accomplishing little except to alert the Germans to an impending attack.[44]

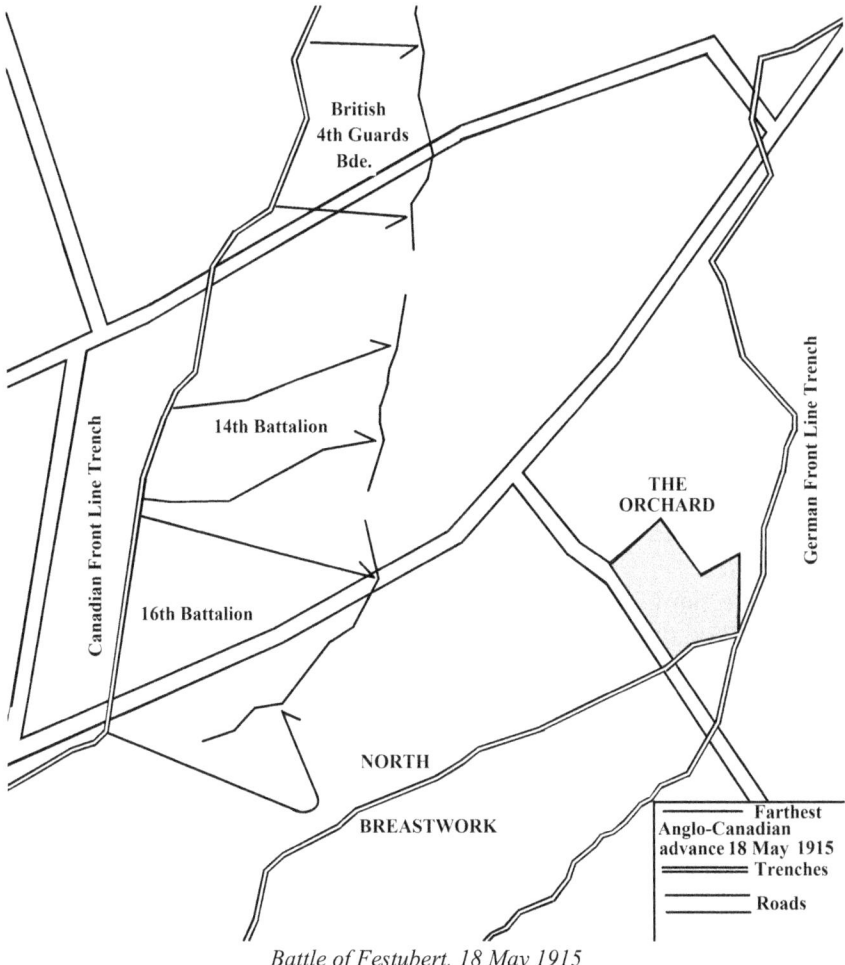

Battle of Festubert, 18 May 1915

Making matters worse, the two hour bombardment scheduled for 2:30 pm began an hour behind schedule. As a result, the Canadians went into the attack an hour late at 5:30, in daylight, after the attack on their flank had stalled, and advanced into the Germans unhindered by the misplaced bombardment. The Canadian assault quickly lost direction in the shell-cratered terrain intersected by waterlogged ditches and stalled after advancing only 400 yards, at which point they had no choice but to entrench.[45] They never got to bayonet-charging distance.

The second Canadian attack at Festubert on 20 May also suffered similar problems. Three battalions (16th, 15th, and 10th from North to South) were ordered to attack along a 1500-yard front at 7:45 p.m. The preparatory bombardment had variable results along the enemy line. In the north it landed on target, but, in the unreconnoitered, south it had less effect. As a result, the 10th Battalion attack on the strong point known as K5 stalled after advancing only 100 yards.[46] In the center, the 15th Battalion advanced against the German front line. The battalion history observed the bounding techniques and the terrain:

> Two hundred yards out and they were advancing grimly against a fire that stiffened with every yard. They were doing it a-la-drill-book – 20 yard rushes, down, up, and on, and down again. They were thinning and there were wide gaps in their line. Touch was lost amid the maze of craters, the tangles of wire, and the scrub trenches come upon here and there.[47]

The 15th Battalion diary gave only scant details of the attack, but the comment the bombardment was "too light"[48] suggests that, here too, the artillery had failed to neutralize German fire. Although the attack went in the under the cover of partial darkness, the light bombardment was sufficient to alert the Germans and the 15th Battalion advance stalled after 300 yards.[49]

In the north, the 16th Battalion continued its advance against the orchard. Having spent an additional two days in the line, the 16th Battalion and supporting artillery were now more familiar with their surroundings as a result of nighttime patrols. With the support of an accurate bombardment and the fire of two machine guns the men of the 16th Battalion reached the orchard and engaged the Germans with the bayonet.[50] Percy Guthrie gave an account of the 16th Battalion attack for the *Montreal Star* a year later:

> On they pressed undaunted, now rushing across open ground, now crawling over a rise, now wallowing up to their hips in a muddy ditch, now wriggling through a hedge strengthened by wire entanglements, and now at last with a mighty yell such as Scots only can utter in times like these, they gain the orchard and with bayonets dispatch the few remaining Huns, the others having fled before glorious stampede of war-mad frenzied men. The attack had followed so closely on the bombardment that the enemy had not time to bring up support to the garrison of the orchard...[51]

A.M. McLennan corroborated Guthrie's account, recalling that they "got the better of [the Germans], before they [the 16th Battalion] realized

[it], they had the orchard. So it paid to follow [the artillery] closely..."[52] The battalion advanced close to the shelling and quickly rushed the German positions as the bombardment lifted.

Battle of Festubert, 20 May 1915

With the orchard secured, Canadian operations focused on K5. On the evening of 21 May, the 10th Battalion attacked again behind the cover of darkness and a three-hour bombardment. The 10th Battalion bombed down a communication trench running between Canadian and German lines and advanced over the open ground on either side of this trench.[53] German machine gun fire checked the left flank of the advance, but well-placed artillery fire and darkness permitted the men on the right flank to assault the German front line. Percy Guthrie described the attack in his narrative:

> Then came the proudest moment of my life. Those brave fellows with cheers that I shall never forget, dashed forward as one man. From the trenches in front came cries of defiance and bursts of flame. The rifles of the enemy barked out their death messages and their machine guns simply rained lead upon us. On the right our boys gained the trench in some places, and the bayonet work began. On the left after the first yell and rush there came a strange silence. The machine guns from K5 were keeping up their infernal stream. I felt our men must be creeping forward silently, but groans from the darkness in every direction filled me with fear.[54]

The 10th Battalion diary also recorded the success of the right flank in "gaining" the enemy front line,[55] but made no comment on the hand-to-hand fighting or the bayonet assault. Guthrie advanced close on the heels of the assaulting waves and became embroiled in two separate hand-to-hand fights in the immediate aftermath of the assault:

> Coming over the trench where there seemed to be something interesting going on, I found myself on top of a Bosche bayonet. I dashed the point aside with my stick and only got a slight touch in the nose, which brought crimson and riled my temper, so I fetched Mr. Hun a whack on the cheek with my stick, what a joke! I have often laughed over it. To think I was so angry at the Bosche that I forgot to shoot him with my revolver, but whoped him with the stick instead...[56]

When entering the enemy trench, or advancing from bay to bay the enemy often suddenly appeared and soldiers were forced to respond with instinct. Here, Guthrie instinctively used the stick to ward the enemy's bayonet thrust and then returned with a swing at his opponent's head. The stress of the situation also saw Guthrie's instincts taking over as he failed to use the revolver in his other hand. After dispatching this foe Guthrie found himself in another hand-to-hand fight moments later:

> I found myself in the grip of a German officer who was vainly trying to get his men to stand up to us. Although smaller, I knew he was stronger than I, but a lucky push from behind upset us and we fell backwards and he tripped on something so that he rolled into a dugout and I happened to be on top, I got a lucky hold in the right spot and the argument was soon over.[57]

It is unclear what "spot" Guthrie had gotten a hold of, but it illustrates that the sudden appearance of the enemy could even prevent the use of weapons readily at hand, like Guthrie's revolver and stick, with hand-to-hand fighting degenerating into a brawl.

The men of the 10th Battalion fended off several German counter-attacks throughout the remainder of the night of 21 May and the next

1915, The Bayonet and Trench Warfare 73

morning. Heavy shelling in the afternoon forced the Canadians to relinquish the captured portion of the German front line, but they retained control of the intersection of the German front line and the communication trench they had attacked down on May 20 and 21. K5 still eluded capture, and the 5th Battalion was brought forward to continue the attack from the intersection held by the 10th Battalion. At 2:30 a.m. on 24 May, as artillery pounded the German positions, the 5th Battalion, supported by elements of the 7th Battalion, closed on K5. Aided by the darkness, the Canadians attacked on both sides of the German front line trench, as well as bombing their way down the German front line trench itself – this time the attack was "successfully driven home."[58] By 2:45 a.m., the two battalions had succeeded in capturing a significant portion of the German front line, and while they were unsuccessful in capturing K5, they managed to make the position untenable for the Germans who withdrew from the redoubt.[59] The Canadians then consolidated the position against German counter-attacks, during which the enemy closed with the bayonet.

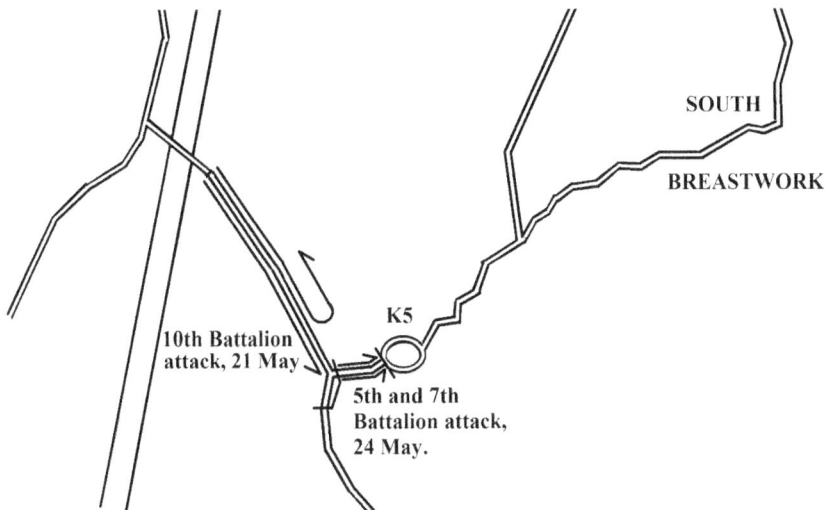

Battle of Festubert, 21-24 May 1915

The official records of the fighting at Festubert failed to observe the particulars of the German counter-attacks that took place at K5; however, two veterans left accounts of the close range of the fighting. Thomas Leo Golden of the 7th Battalion took part in a counter-charge against a German bombing party: "While we were here [Festubert] a party of Germans came along with bombs to try and throw at us. We fixed bayonets and started to climb over the parapets. Just as soon as they saw the steel they ran as fast as their legs could take them."[60] The threat of the bayonet charge demonstrated both the significance of the negotiation of close combat, and the range (roughly that of hand grenades) at which the negotiation took place. Sergeant Harold Baldwin had an even closer encounter in the hand-to-hand fighting that took place at Festubert. Wounded in the leg during the consolidation, Baldwin was saved by the bayonet work of one of his comrades during one of the German counter-attacks that morning:

> I had merely commenced to feel the sting of the pain when the Huns rushed us again and it was hand-to-hand. A Bavarian lunged towards me with rifle clubbed; I closed my eyes, as I was utterly helpless and waited for my skull to be smashed. The blow did not fall. I opened my eyes just in time to see our sergeant-major plunge his bayonet through the Bavarian's neck. Down flopped the Hun on all fours, with his hands one on each side convulsively clutching at the bayonet."[61]

Official records proved incapable of recording the close combat that individual soldiers witnessed and participated in. There was simply not the time or space to write down these details of battle.

The final Canadian attack at Festubert, which illustrated the close association between bomb and bayonet, took place on the night of 24 May. Two small parties – one each from the 3rd and 4th Battalion – were to advance from the orchard while the artillery bombarded enemy positions at 11:00 p.m. The objective of the attack was to enter the German line and "with bombs and bayonets clear the trench."[62] Each party was made up of 10 bombers and 10 bayonet men. In addition to the darkness and the artillery bombardment, they were also covered by a ruined house in No Man's Land. These teams were to bomb and bayonet their way down a 600-yard length of the German front line, at which point reserve troops were to advance and consolidate.[63] Due to confusion in the interpretation of orders, the barrage and the attack were delayed for thirty minutes. The artillery succeeded in cutting the German wire, but it failed to silence the German machine guns. Only one of the parties succeeded in entering the enemy trench at 11:40; the attacking force was then checked by German machine gun fire. Only twenty minutes after starting, the attack was cancelled.

The fighting at Festubert demonstrated several factors that could impede an attack before the infantry could assault the enemy position with the bayonet. The failure of artillery to sufficiently neutralize enemy positions featured prominently in the after-action reports; however, failure can be attributed to several factors. First, the chronic shell shortages led to lighter bombardments that were ineffective in neutralizing enemy fire. Second, the artillery was also still mastering the new techniques of firing at targets beyond their line of sight, a problem exacerbated by the shell shortage as it presented artillerymen with diminished opportunities for practice. Finally, the artillery was hampered by poor cartography which made indirect fire largely a matter of luck and guess work. The rapid pace of the operations, late bombardments, and uncoordinated attacks demonstrated problems on the part of staffs and commanders in planning set-piece battles. Until the problems of intelligence, planning, and cooperation could be overcome, the infantry had extreme difficulty in closing with the enemy and assaulting with the bayonet.

V. Givenchy: New Tactics and Unforeseen Consequences

The Canadian Division's major attack at Givenchy encountered additional problems for the infantry finding their way across No Man's Land and assaulting trenches with bomb and bayonet. The 1st Battalion was ordered to attack near Givenchy, 2 kilometers south of K5, supporting the right flank of a larger operation by the British 7th and 51st Divisions. At 6:00 p.m. on 15 June, 1915, the attack went in after a half hour bombardment on the German front lines. Just before the advance, 3000 pounds of explosives were detonated under the German front line trench in the center of the battalion's advance.[64] The mine explosion seriously damaged the Canadian jumping-off trenches, causing 50 casualties amongst the infantry and bombers on the right flank, including the bombing officer, and burying their reserve supplies of grenades.[65] The left bombing parties also lost their grenade reserve, which was detonated by the force of the blast.[66] This loss of grenade reserves had serious repercussions for the attack.

On the left, the bombardment and the mine were insufficient to neutralize the enemy, and machine gun and rifle fire checked the advance of No. 2 and No. 3 Companies. On the right, where the soldiers followed closely on the heels of the bombardment, soldiers of No. 4 Company and No. 1 Company crossed the open ground quickly and entered the enemy front line at the south lip of the mine crater. After entering the German line, three bombing parties, consisting "each of 7 grenadiers (one of six only) [and] 3 bayonet men,"[67] proceeded to bomb and bayonet their way toward the objectives. Shortly after 7:00 p.m., the Germans responded by "counter-bombing"[68] the Canadians, meeting offensive action with offensive counter action in an intense and close range application of *élan*: "Violent hand-to-hand fighting and bombing ensued until owing to our supply of bombs having become exhausted and our numbers having been so depleted through losses, our forward position became untenable..."[69] Without bombs and reinforcements, the Canadians were unable to push forward against German bombing parties with only the bayonet – as a result, they were eventually forced back.

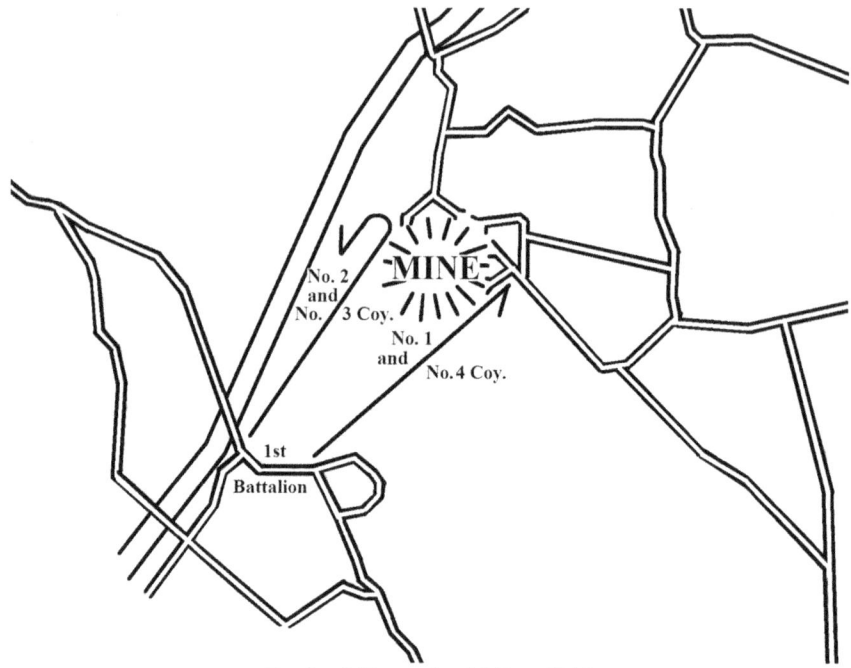

Battle of Givenchy, 15 June 1915

Coordination of infantry and artillery had improved somewhat with the Canadian Division's attack at Givenchy. However, the unintended damage from the use of mining showed how costly the learning process involved in the assimilation of new weapons and tactics could be, and demonstrated the impact a planning error could have on the success of an attack. Givenchy also proved the importance of supporting an attack after it had gained the enemy trench – without adequate reinforcement and suppression of enemy fire once a position was taken, German counter-attacks would successfully drive home with bomb and bayonet to reclaim lost positions.

Canadian offensive operations in 1915 had demonstrated that, given proper supporting fire to suppress the enemy, the infantry could cross No Man's Land and assault enemy positions with the bayonet to gain a toehold in the enemy line. Once in the enemy trenches the combination of bomb and bayonet permitted attacking troops to exploit the initial breach. The failure of Canadian attacks was not the result of a deficiency in the bayonet; rather, failure emanated from two basic problems. First, if artillery was insufficient in neutralizing enemy fire, the enemy's small arms and artillery fire would break up the attack before the bayonet assault could be launched. Second, if reinforcements, machine guns, and stores – grenades in particular – were not moved forward quickly enough or in sufficient quantities, the assaulting forces were driven from their gains by the inevitable counter-attacks.[70]

VI. Bayonet Training 1915: Formal and Informal

Throughout 1915, infantry training suffered as a result of the rapid expansion of armies on the Western Front. In order to deal with the deluge of recruits and the needs of hemorrhaging of manpower on the Western Front, the basic training syllabus was shortened in the summer of 1915 from 24 weeks to 10 weeks. Within this new syllabus, bayonet training dropped from 20 hours to a scant nine hours.[71] However, front line soldiers found that neither the twenty hours of training – much less nine – nor the techniques contained in the official systems of bayonet fighting were sufficient for their needs. Soldiers, therefore, developed unofficial techniques which were propagated by informal training; this informal training system proved highly effective and unofficial techniques even began to permeate official training instruction. As well, the publication of unofficial training literature saw a resurgence in England as soldiers sought to fill this front-line need.[72] Towards the end of the year, bayonet fighting deficiencies received chain-of-command attention. While infantry-training syllabus was still 10 weeks, the bayonet training within the syllabus was increased to 24 hours.[73] In October 1915, foundations were laid by the Canadian Training Establishment in England for the establishment of an administration responsible for overseeing bayonet training.

The fusion of formal and informal techniques was captured in a poem written by an anonymous member of the 64th Battalion, entitled "Do you remember bayonet fighting in 1915?"

> When I says 'Fix' you don't fix,
> Just grip bay'nets and wait,
> Extend your rifles out a bit,
> Eyes right, then hesitate;
> When I says 'By-nets', everyone
> Will move in unity,
> You whips 'em out and 'wops' 'em on.
> It's simple, as can be;
> Then watching close the right-hand guide
> Three paces out in front,
> You brings your arms back to your side,
> Which ends this little stunt.
> 'In, Out – On Guard' the sergeant said,
> He barked it loud and shrill,

The new recruit just shook his head,
What is this funny drill?
'Right Parry – point' the next command,
It rang across the square,
'That's not a dung fork in your hand –
Wake up that man back there'
'You'll not be playing croquet now,
You're here to learn to kill,
My job it is to show you how,
And show you how, I will'
'I want to see you all get sore,
Get murder in your eye,
Remember that you're out for gore,
The enemy must die.'
'The bayonet is a useful thing
Some day your life may save,
So let me see you use some 'zing,'
And let me hear you rave;
'Left Parry and Butt to the chin,'
And don't forget your knee,
There's places you can throw it in
Most devastatingly'
'Now, see those bags hung over there?
Each one contains a Hun
For charging now you will prepare,
And take it on the run;'
'High Port' your arms and off you go,
Let's hear you curse a bit,
You're out to get the Hun, you know,
So put some 'zip in it'
'You get him first or he'll get you,
And though you curse me now,
The day will come when you'll not rue
The time I showed you how.'[74]

The pattern of training that had evolved by the end of 1914 is clearly in evidence here, with recruits practicing their postures in air and then moving onto the assault course. The techniques advocated in *Infantry Training 1914*, with its simplified *long point* and parries, as well the *infighting* technique of the *butt stroke*, are present too. However, two informal techniques appear in the poem, the knee and the killing face. The use of the knee was not advocated in official training literature until 1916. The killing face is suggested by the lines "I want to see you all get sore, / Get murder in your eye, / Remember that you're out for gore." The killing face, as discussed in Chapter Three, was meant to aid in intimidating the enemy.[75]

The final lines provide the context and starting place for this examination of bayonet training in 1915 'And though you curse me now, / the day will come when you'll not rue / the time I showed you how.' Soldiers came to have respect for this 'funny drill.'

Front line soldiers found the bayonet a critical tool in trench warfare. Another poem appearing in the January 1916 edition of *The Brazier*, a Canadian trench newspaper, demonstrated the importance of the bayonet to not only the regular infantryman, but for the specialist sniper as well:

> What makes the sniper's heart to break, what makes him to perspire?
> It isn't carrying sacks of coal to stoke his dugout fire;
> It isn't packing leather coats and other airy trifles
> Like sheepskins, blankets, water-proofs, it's humping around two damned rifles.
>
> Oh! The telescopic rifle with its telescopic sight
> For telescopic slaughter may be perfectly all right;
> But the sniper quietly finds; that it's a blessing somewhat mixed,
> When he has to hump another gun who's bayonet can be fixed.[76]

The sniper's preference for the Ross rifle has been well documented,[77] but as the Lee-Enfield came into Canadian service in greater numbers, Ross bayonets became more difficult to find. Even if the sniper could find an appropriate bayonet, the fragility of telescopic equipment made such rifles unsuitable for bayonet fighting. In addition to changing the balance of a rifle, the fixed bayonet changed barrel phonics and muzzle pressure of a firing rifle, affecting rifle accuracy at longer ranges. Regardless, the observation that a sniper needed a rifle to fix a bayonet upon testifies to the soldier's reverence for the bayonet as a weapon.

Nor was this strictly a habit of snipers. F. G. Bagshaw, with the 5th Battalion, recalled the tendency of Canadian troops who scrounged Lee-Enfield rifles to keep their Ross rifles in 1915. "We all had two rifles when we came out [of the line at Ypres], the Ross and the Lee-Enfield. The reason we had to carry two was because our bayonets wouldn't fit on the Lee-Enfield, you see."[78] The problem for these soldiers in 1915 was exactly the opposite of that which faced snipers a year later; it was easy enough to snatch a Lee-Enfield left unattended by a British soldier. But acquiring a bayonet, which was worn on the belt, often presented a more difficult proposition. The fact that soldiers informally increased the weight of their equipment by 10 pounds or more for the sake of a fixed bayonet speaks to the infantryman's fondness for the bayonet in battle.

Given the growing importance of the bayonet to the infantryman in the trenches and the strains on the system of bayonet training, informal training and unofficial techniques evolved in the front lines. Harold Baldwin's memoir of the war gives an account of one of these informal training sessions as the First Contingent acclimatized to trench warfare in the spring of 1915:

> It was while trying to keep warm that first night over the little charcoal fire that I first learned how to handle my bayonet, if I was ever to be lucky enough to ram it so far into a German belly that I couldn't pull it out handily. The lesson came from a corporal of the East Lanks (Lancashires) who was explaining the advantages of the Lee-Enfield rifle and bayonet over the Ross, and his description was so realistically vivid that my teeth forgot to chatter with the chill I had.
>
> 'You see,' he said, 'if you push it in too far, you canna get it oot again, because this groove on the side o' it makes the 'ole air-tight; as soon as it is jabbed into a man the suction pulls the flesh all over it and you canna chuck it oot.'
>
> 'Well, what would you do if you couldn't get it out and another mug was making for you?' I asked.
>
> 'Why if a twist won't do it, stick your foot on the beggar and wrench it out; if that won't do it, just pull the trigger a couple of times and there you are – she will blow out.'
>
> 'Did you ever have any trouble yourselves?'
>
> 'Oh, aye. I remember at Landrecies, in the 'ouse to 'ouse fightin', my chum, Topper, and me were backed into an alley, with a wall at our back and a bunch of hulking Prussians pressing us hard. Some more of the boys fell on them hrom the side, but Topper and me had all we could do with the two or three that took a fancy to us. The Pruss that took a fancy to me raised the butt of his gun to smash me nut and I took a chance an lunged. I lunged too 'rd and I 'ad the trouble I've just been telling ye, and in my funk I did just what I told ye; I twisted – she stuck; I wrenched and tugged – she stuck; and if I 'adn't fired and got the bloomin' blade free, I wouldn't be 'ere a-tellin' yer about it.'
>
> 'And why couldn't I do the same with this one?' I asked, referring to my Ross bayonet.
>
> 'It's too broad at the point. The man that gave ye that dam'd thing might just as well 'ave passed sentence o' death on yer in a 'and to 'and go.'[79]

All three of the techniques described – the twist, bracing with the foot, and firing a round – to assist in the withdrawal of the bayonet appear in the accounts of front line soldiers,[80] yet only planting the foot and firing a round eventually appeared in official training literature during the war.[81] While informal training proved effective at disseminating additional techniques of bayonet fighting, it provided little guarantee of uniformity.

As observed in "Do you remember bayonet fighting in 1915?," the trenches were not the only place these unofficial techniques were taught. The second year of the war also saw a resurgence of privately published training literature that advocated the use of unofficial techniques. *Practical Bayonet Fighting: With Service Rifle and Bayonet*, for example, advocated the use of the knee and boot in *infighting* a year before it was advocated in official training literature.[82] This private manual also included an elaborated syllabus for bayonet training.

Unofficial techniques began to penetrate the training camps and the lessons of the Army Gymnastic Staff (AGS) instructors themselves. David MacMillan, a member of the Second Canadian Contingent, recalled an Imperial AGS instructor teaching Canadian assistant instructors to twist their bayonets:

> The equipment used was quite simple. There was a gibbet-like structure with four dummies of straw. The troops advanced on these dummies at the run, rammed the bayonet in, gave it a jerk and pulled it out. The instructor was also a Sergeant in an English Regiment, and when making the charge he would exhort us 'to hadopt ae fierce hexpression, and be sure to give the bayonet a good twist before wifdroeing.' This was supposed to have some effect on the victim, but what, I never did find out. Probably it was to ensure that the bugger would never again 'Hoch der Kaiser.'[83]

In addition to the twist of the bayonet – which, in this case, was not necessarily to aid in the extraction of the weapon – this instructor also advocated the use of the "killing face," which was still an unofficial technique in 1915.

McMillan also recalled disinterest as another potential problem gripped the training system in 1914 and 1915:

> On my return to the regiment, I was detailed with several others to instruct in bayonet fighting, and I will not dwell upon this assignment except to say that I shouted 'in' 'out' 'on guard' so often that my voice was hoarse and my throat dry at the end of each session. The remedy, however, was close at hand – a couple of pints of beer in the Mess and I was back to normal.[84]

As observed previously the loss of bayonet fencing from training during 1914 had the effect of making training and techniques stiff and robotic, the simple repetitious routine of words of command by impressed NCO's could easily lead to further stiffness and disinterest in the training. Nor was McMillan alone in the observation of soldiers not taking their training in bayonet fighting seriously. *Practical Bayonet Fighting...* warned instructors to observe the manner in which trainees extracted their bayonets from bayonet sacks lain on the ground to practice bayoneting entrenched opponents:

> In the final assault avoid the tendency of 'haymaking' when pointing at the sacks. The bayonet should be thrust into the sack and withdrawn properly and not released by hoisting the sack over the shoulder. It is very difficult and quite unnecessary to cast your opponent over your shoulder like a sheaf of corn on a fork.[85]

As noted previously, the system of AGS instructors training a four-man cadre for each battalion permitted the rapid training of men, but this left the control of training up to the battalions and the whims of their assistant instructors. While fostering innovation in the face of an inadequate official training system, the combination of informal techniques and unofficial literature prevented uniformity in training and created a chaotic situation governing a weapon that was proving important on the Western Front.

Towards the close of 1915, the General Staff began to consider how to regain control of the situation, likely due to the impending increase of bayonet training in accordance with the November 1915 syllabus. In September, Major Henry George Mayes* suggested the establishment for the Canadian Army Gymnastic Staff (CAGS) in the Canadian training camps in England for training battalion instructors in bayonet fighting and physical training.[86] Mayes continued to develop his proposal and arranged for one officer and four NCOs from the Canadian training establishment in England to attend the AGS bayonet instructor's course at Aldershot in October and November. Additionally, Mayes secured the promise of instructors from the AGS to assist the establishment of a Canadian bayonet fighting school at Shorncliffe.[87] Finally, in order to unburden the now-overwhelmed bayonet training system in England, Mayes further proposed the establishment of bayonet fighting schools in Canada.[88] Mayes' proposals found a receptive audience – bayonet-fighting schools were established in Ottawa, Winnipeg, and Montreal by the end of 1915.[89]

---

* Henry George Mayes would become the Director and Inspector for Bayonet Fighting and Physical Training in 1916, Director of the Canadian Army Gymnastic Staff in 1917, and advisor on physical training for the newly established Royal Air Force in 1918. The following biography of Lt. Colonel Mayes, dated Sept 30th, 1918, appears in RG 9, III-D-1, vol. 4178, Folder 113, file 13: "Lieut. Col. H. G. Mayes, Director of the Canadian Army Gymnastic Staff, has been appointed advisor to the Air Ministry on the physical and athletic training of the Royal Air Force. Col. Mayes, whose manual of bayonet fighting has been issued under the authority of the War Office has largely contributed to the fine fighting qualities of the Canadian Force, his theory being that no man can be an efficient soldier unless thoroughly fit and that courage largely comes from confidence in fighting weapons. A believer in the 'Spirit of the Bayonet' by the constant use and practice of that weapon he brings his men into condition by every kind of sport, notably boxing, as being the best means of cultivating the combative idea."

Heightened official interest in bayonet fighting yielded a new official training pamphlet by the end of 1915. *Bayonet Fighting: Instruction with Service Rifle and Bayonet: 1915* was built on the foundations of *Infantry Training 1911*, although it provided some refinements. Included was the new *infighting* technique of the *jab*, a close attack delivered by moving the right hand forward to the bayonet mounting and then stabbing the adversary under the chin. One of themes evident in British bayonet fighting between 1905 and 1918 was the development of techniques for fighting at progressively-closer ranges; the *jab* was part of this evolution.

Figure 8: "The jab" (Bayonet Fighting 1916)

In addition to the *jab*, the 1915 pamphlet also suggested the use of a rolled up newspaper fastened to the end of a stick in order to provide a manipulated target for butt strokes and parries, called a "parrying stick" or "blob stick." The development of the "blob stick" itself was a response to the inadequacies in bayonet training caused by the loss of "loose play" at the end of 1914. Instructors invented informal methods of teaching the techniques of the butt stroke and the parry, but many of these informal methods risked damage to rifles and bayonets. In addition to bringing uniformity to bayonet training, the new training administration was also intended to stem the wastage of equipment from unofficial training experiments. For example, *Bayonet Training 1916* emphasized, "The greatest care should be taken that the object representing the opponent and its support should be incapable of injuring the bayonet or butt, and only light stick must be used for parrying practice."[90] The slow development of the blob stick in 1915 and 1916 was a sign of the inability of official training systems to control informal invention. However, with the establishment of an official training administration in 1916, the "blob stick" became a standardized training tool and the solution to some of the weaknesses in training since the loss of "loose play" in 1914.[91]

By the end of 1915, the strains on the system for training soldiers in bayonet fighting had become painfully evident, but so too had the importance of the bayonet in battle. These problems, which had caused the intrusion of informal techniques into official training, were addressed officially towards the end of 1915, but the struggle to regain official control over training persisted well into 1916.

VI. Conclusion

Front line soldiers were quick to realize the tools and techniques required in trench warfare even before the end of 1914. Improvised grenades and mortars had been developed in the first months of the war. In the summer of 1915, the rifle grenade and the Lewis gun joined these new weapons. The assimilation of these weapons, like the debate between *dispersion* and *élan* in the decades before the war, demonstrated that armies were not inflexible. As the armies rapidly adopted new weapons in response to trench warfare, the bayonet was retained for the same reason the grenade and the mortar were adopted – it was seen as an important and useful weapon.

The failure of many attacks in 1915 were in large part due to the problems of neutralizing enemy fire and then reinforcing the attacking infantry, while integrating new tactics and technology. For example, an issue like poor mapmaking had a profound impact on Canadian operations at Festubert, the result of which was artillery firing off target and accomplishing little but alerting the Germans to an impending attack. At Givenchy, the revived use of medieval mining techniques devastated the German front line, but damaged the Canadian line as well. Above all, the inability to destroy or subdue enemy machine guns and artillery meant that the advancing infantry had difficulty closing to bayonet-charging range, whether they advanced in dispersed formations or not.

The Canadian operations of 1915 demonstrated that if properly supported with machine gun or artillery fire to neutralize enemy weapon systems, infantry could clear enemy positions with the bayonet in the initial assault. These operations also demonstrated that, if supplies of munitions and reinforcements could be maintained, Canadian troops were capable of holding onto their gains through cooperative use of the hand grenade and bayonet. Front line soldiers quickly realized the bayonet's significance and improvised techniques and training until institutional friction could be overcome. Even before the end of 1915 it was realized within both official and unofficial circles that the place of the bayonet had been underestimated, the challenge of 1916 would be that of creating uniformity and consistency out of the initial chaotic responses.

Chapter V: 1916, The Bayonet and the Battle of Materiel

THE ELEMENTARY PROBLEM of supporting the attacking infantry during 1915 persisted into 1916. Tactics were refined over the course of 1916, but there were still unforeseen consequences and additional problems. Further complicating matters was the ongoing expansion of Canadian forces in France, whose new formations often had to learn the lessons of 1915 through hard experience. In spite of the on-going evolution of the set-piece battle, the principles of the attack remained the same: to neutralize enemy machine guns and artillery so that the infantry could cross No Man's Land. If sufficient weight of attacking infantry could get within 30 yards of the enemy trenches, the bayonet proved effective at forcing the negotiation of close combat and clearing enemy positions. Learning how to get the infantry to this range was slow, costly, and fraught with setbacks. But the increasing frequency of hand-to-hand fighting attested to the Canadian Army's progress in bringing the infantry into bayonet range.

I. The Problem of the Offensive

The primary problem of the attack in trench warfare was the open killing ground of No Man's Land, and many Canadian assaults were broken up well before they came to bayonet range. Without proper supporting fire from artillery, or cover from terrain or darkness, the attacking infantry was caught in the open spaces of No Man's Land, stuck in uncut wire, cut down by small arms and machine gun fire from unsuppressed positions, or killed by defensive artillery fire. After a successful attack, unsuppressed enemy defensive fire could and did harass or pin down the movement of support troops and stores across No Man's Land, and without sufficient support, attacking troops were driven from their gains.[1]

If Canadian infantry could cross No Man's Land in sufficient numbers and come to bayonet range (between thirty and fifty yards), the final rush rarely failed to enter the enemy trench. This point was emphasized by Lieutenant J. S. Williams in a letter home in April 1916: "Everyone who has been out here is thoroughly acquainted with the fact that it is one of the easiest tasks in this war to *capture* a trench but to *hold* on to it is something totally different."[2] Williams glossed over the grim reality of the assault and the struggles with bomb and bayonet to clear and then hold a captured trench. However, once in the enemy trenches, attacking troops were frequently successful in exploiting the breach in the enemy lines, relying on hand grenades before rushing the surviving occupants in the next bay with the bayonet.

Once the position was captured, reinforcing troops and stores were rushed across No Man's Land and the process of consolidation began. Enemy counter-attacks overland were to be broken up by small arms fire and prearranged defensive artillery fire. Against bombing counter-attacks, bomb and bayonet tactics were critical for stalling German offensive momentum. Both sides pushed forward throwing bombs and charging home with the bayonet in the seesaw fighting that occurred around hastily erected trench blocks. Few Canadian attacks were broken up during the initial bayonet assault; even formations that had suffered severely while crossing No Man's Land frequently carried enemy positions with a bayonet charge. The problems arose when an attack ran out of forward momentum and the attacking troops were forced onto the defensive.

One solution was to increase the scale of operations and throw more resources into the attack: more guns, shells, and soldiers.[3] The Germans named this approach to war the "battle of materiel," or *Materialschlacht*.[4] The shortage of artillery ammunition that had plagued Entente operations in 1915 was partially remedied in 1916; however, high explosive (HE), which proved far more effective against entrenched defenders and barbed wire, was still not produced in the quantities required for trench warfare. The scarcity of HE was remedied over the course of 1916, but the mass production of shells and fuses for artillery presented additional problems.[5] The artillery-intensive tactics of the Somme, for instance, suffered from a high proportion of dud shells.[6] HE fuses also often failed to detonate in the mud and shells buried themselves deep in the ground before exploding.[7] These problems hampered the reduction of wire entanglements and fortifications.[8] While artillery tactics improved over the course of 1916, sufficient friction still existed to make the task of getting the infantry within bayonet range uncertain.

II. St. Eloi Craters and Mount Sorrel: New Divisions, New Corps, New Problems

The Canadian battles during the first half of 1916 suffered from additional complexities in the conduct of offensive operations. The poor showing of the Canadian 2nd Division at the St. Eloi craters in April 1916 can be attributed in part to inexperience among commanders and staffs, who were still learning their business in difficult conditions.[9] The terrain at St. Eloi was shattered by weeks of previous fighting and the detonation on 17 mines in British operations to take the area had destroyed all land marks, exacerbating matters was the rain which turned the broken ground into a morass of mud and prevented aerial observation of the fighting. The conditions would have been trying enough for experienced formations, but St. Eloi was the first major operation conducted by the 2nd Division, which had arrived in France in September 1915. It was also the first major operation undertaken by the newly established Canadian Corps Headquarters, which had been established upon arrival of the 2nd Canadian Division. The Canadians took over the line (ahead of schedule) from British units mauled by the fighting for the craters, and thus did not have the opportunity to reconnoiter the terrain. The Canadians had only just relieved British units in the line when the German counter attack struck on the night of April 5th, 1916, and, threw Canadian troops out of their unfamiliar positions. 2nd Division counter attacks became uncoordinated and lost in the broken night time terrain. The fighting at St. Eloi was also hindered by inadequate intelligence, and Canadian commanders experienced great difficulty in locating Canadian troops in the mire of mud and craters, much less the enemy.[10] Commanders seemed at a total loss for how to gain control of the situation. Under such conditions, the coordination of infantry and artillery was impossible.[11] The 2nd Division had suffered from deficiencies in intelligence and navigation in the shell blasted environment of No Man's Land in a similar fashion to the 1st Division's first set-piece attacks at Festubert had in May 1915.

 The debut of the 3rd Canadian Division at Mount Sorrel on 2 and 3 June was similar to the 2nd Canadian Division's experience at the St. Eloi

Craters. On 2 June, the Germans unleashed a heavy bombardment on the positions held by the 8th Brigade (PPCLI, 1st CMR, and 4th CMR from north to south) north of Mount Sorrel. The intense shellfire shattered the Canadian trenches and front line battalions causing considerable casualties, including both the divisional and brigade commanders.[12] After the barrage, the Germans exploded four mines under the battered Canadian positions before they surged forward and overran the remnants of the two battalions of the CMR. On the north flank of the German advance, the PPCLI, in spite of heavy casualties, prevented the Germans from widening the gap in the Canadian lines. Only scant records remain for the three defending battalions and only the PPCLI history makes direct reference to bayonet fighting in the defence of Mount Sorrel.[13] Undoubtedly, much of history of the defence of Mount Sorrel was lost in the high casualties among the defending formations: the PPCLI suffered more than 400 casualties, the 1st CMR suffered 80% casualties, and the 4th CMR suffered 89% casualties.[14]

A nighttime counterattack was immediately ordered for 2:00 a.m. on 3 June. The attack became confused by the fluid front lines. For the beheaded 3rd Division committing its troops into a major attack for the first time, the coordination and planning for the attack was haphazard at best. The attack was postponed several times and it was daylight before the attacks began. In the face of such confusion, the uncoordinated Canadian attacks generally faltered well before the German positions could be assaulted; however, some Canadian troops did succeed in entering German positions and engaging in hand to hand fighting.[15] In the 3 June counter-attack at Mount Sorrel, inadequate planning and coordination had resulted in failure.[16] After the failure of the first counter-attack at Mount Sorrel, the Canadian Corps adopted a more deliberate approach to regaining lost ground. On 9 June, artillery began a four-day bombardment of German positions. During the bombardment, the artillery executed four feints in order to wear down and confuse the defenders as to the main point of the attack. These feints involved increasing the intensity of the bombardment and then lifting fire as if in prelude to the infantry assault. The Germans were deceived into standing-to in order to repel assaults that did not come – instead the bombardment was resumed and the Germans were again forced to ground.[17]

At 1:30 a.m. on the morning of 13 June, the Canadian attack began. The artillery bombardment intensified and then lifted off the front line German positions for a fifth time. Just before it lifted, three battalions of the 1st Division (3rd, 16th, and 13th) advanced. On the left flank of the main advance, the 58th Battalion bombed its way down the communication trenches toward the old Canadian front. To the southwest, the 1st Battalion moved overland to secure the right flank. The final objective was the original Canadian front line. The 3rd Battalion issued a no-firing order to the troops, meaning they would plunge into the German lines bayonet-first. Even for the battalions that did not issue a no-firing order, the mud churned up by two weeks of fighting had the same effect, as "men fell down in the mud and were plastered with it; revolvers and rifles became coated with it, and were rendered useless."[18] The no-firing order and the battalion history noting the peculiar situation of soldiers not being able to fire in the attack due to the mud, demonstrated that the men of the 1st Division were accustomed to firing on enemy positions as they moved forward in small groups, a reflection of the fire and movement tactics advocated in prewar training.[19] Under a substantial weight of artillery fire and the cover of darkness the infantry stealthily closed to bayonet range.

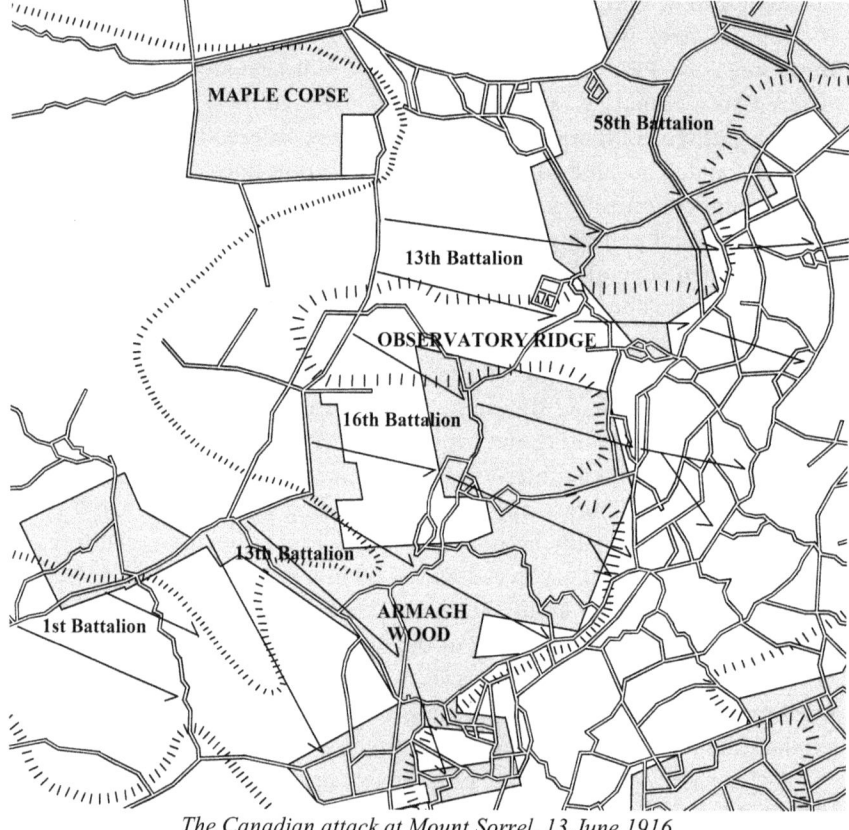

The Canadian attack at Mount Sorrel, 13 June 1916

At the German front line, the assaulting battalions surprised the defending troops, who were only able to offer haphazard resistance with rifle and machine gun fire. The 16th Battalion history provided a detailed account of the assault of the front line:

Numbers 1 and 2 Companies scrambled toward the 'Halifax' objective as best they could... The enemy opened up a short burst of rapid fire from their front trench, but it was ineffective; they then threw their bombs and the 16th men retaliated with bombs, for these weapons survived the mud; and then went in with the bayonet, at the sight of which the resistance collapsed. Many of the Germans appeared dazed. Most of them had no rifle and had no equipment on, and those who had, made no pretence of further fight once they saw the steel.[20]

For the 16th Battalion attack, the weight of the artillery bombardment and the sudden appearance of bayonet-charging Canadian troops forced the defending Germans to surrender. In other sections of the German front line, defenders chose to resist the bayonet charge, and this led to fierce hand-to-hand fighting.[21] The 3rd Battalion, for instance, "met little opposition in the [front line] trench and bayoneted the Germans in the trench."[22] The individual perspective differed from that of the battalion diary – Lieutenant H. R. Alley, of the 3rd Battalion recalled this incident of "little opposition" as "some pretty bloody bayoneting and the butting, if you were too close for the bayonet."[23] In the front line trench, and in the two trenches beyond, the bayonet work continued after the initial assault as bombing parties fanned out and secured the trenches.[24]

In accordance with the artillery schedule, the assaulting battalions had ten minutes to capture each trench and then move on to the next.[25] At 1:40 a.m., the bombardment lifted off the second German trench line to the original Canadian front line and the assaulting battalions advanced. At the second line German resistance stiffened and they were more willing to engage the Canadians in close combat, as a result there were an increasing number of Germans put to the bayonet. For example, the 3rd Battalion diary records that the Canadians "carried the trench and bayoneted most of the occupants."[26]

Overcoming the increased resistance at the second line also included small groups and individuals flanking and storming isolated machine guns.[27] One attack on a lone machine gun amongst the charred stumps of Armagh Wood demonstrated the chaos of close combat, in which the bayonet charge was also organized by small groups of men, or even individuals, to overcome strong points and resistance as they encountered them:

The flame from the gun could be seen in the midst of a black jumble of fallen trees; a rush was made toward it but with no success. Thereupon Bell-Irving, discovering a clearing on the left, dashed ahead and coming on the gun crew from a flank, disposed of them single handed... from the parapet of the German machine gun emplacement, bayoneted three of the gunners, and was lunging at a fourth when an opponent grasped and held the rifle. By this time, however, some Battalion men had come up and Bell-Irving, very tired of bayonet exercise, left the rifle in the German's hands. He asked the newcomers to attend to the matter, and securing another bayonet, went forward.[28]

Improved all-arms cooperation yielded results as well. At 1:50 a.m., the bombardment shifted to the old German front line to protect the newly won positions against anticipated German counter-attacks.[29] The three leading battalions pushed forward and assaulted the final trench. German resistance began to falter in the face of the Canadian advance: "the crest was carried with slight loss," recorded the 3rd Battalion diarist, "many Germans being bayoneted before they could get away."[30] All three responses available to defenders in the negotiation of close combat were demonstrated at Mount Sorrel: fight, flight, and surrender.

The fighting at the St. Eloi Craters and the 3 June attack at Mount Sorrel drove home to the Canadian Corps the importance of accurate intelligence, competent staff planning, and coordination of attacks. Without these things, bringing the infantry into bayonet range of the enemy was impossible due to German fire, or pointless due to the impossibility of resisting German counter attacks. Both careful planning and use of overwhelming artillery support became hallmarks of successful Canadian set-piece operations during 1917. The carefully planned bombardment and the cover of darkness had permitted Canadian infantry to assault the enemy positions with the bayonet.[31] Yet in spite of these refinements and their influence on future operations, the broad prewar principles of the infantry attack were still evident.

III. Interlude: The Lessons of the Somme

The pressure to expand the British Army and the need to replace casualties had led to an increasingly shortened training syllabus. The pre-war 24 week schedule had shrunk to as little as nine weeks in the middle of 1915, but, had been increased to 14 weeks by the end of 1915.[32] The British Army now had sufficient manpower, but its qualities as a fighting force were questioned within the chain-of-command due to the reduced training regimen.[33] Artillery ammunition had also become plentiful, as the shell crisis of 1914 and 1915 was resolved. In response to these conditions, tactics were simplified for the attack on the Somme – infantry, heavily laden with the equipment they needed for the clearing and holding of trenches, were to advance in easily-controlled tight formations to occupy German positions battered by a week-long bombardment.[34] These tactics were a logical decision on the part of the General Staff: to lead with their strength in artillery and downplay the weakness of the infantry.[35] However, much to the disappointment of the staff planners, artillery ammunition, like manpower, suffered from deficiencies in quality.[36] The fighting on the Somme also revealed the strength of German defences and the requirement for well-trained infantry to close with the enemy.

The scale of the 1 July operation, with 120,000 soldiers advancing into battle, forced a tightly scheduled plan controlled by the highest levels of command with little leeway for variation. The failings of the now-plentiful artillery ammunition became apparent as dud shells and faulty fuses failed to cut wire or sufficiently damage German trenches. When the bombardment lifted, the attacking infantry left their trenches and formed into waves under the watchful eye of the Germans. The British infantry were channeled into tight killing zones by the wire and cut down by German artillery and small arms fire.

In the face of nearly 60,000 casualties on the first day of the Somme, a reappraisal of the British tactics of 1 July was unavoidable. The second half of 1916 saw a flurry of memos on lessons learned at the Somme.[37] For the infantry, this resulted in renewed emphasis on the platoon in the attack and the devolved control of operations. Commanders and staff also attempted to shorten the pause between artillery fire and infantry movement in the attack. Experienced units, such as the 16th Battalion at Festubert, had experimented with advancing close behind the bombardment during 1915 and 1916. This technique of "leaning on the barrage" became standardized practice in Anglo-Canadian forces.[38] Attacking infantry were to advance to within 50 yards of the barrage, and then assault the enemy position with bayonet and bomb as soon as the barrage lifted.[39]

The nature of the bombardment also began to change. As seen at Mount Sorrel, the artillery lifted from one German trench line to batter the next. This left German outposts in front or behind the fall of shot unsuppressed by the artillery. In response to this deficiency, the creeping barrage was developed in the summer of 1916.[40] The creeping barrage began forward of the enemy position to be assaulted and then lifted a set distance (typically 100 yards) every pre-determined number of minutes (usually 3 minutes); while the infantry moved to within 50 yards of the falling rounds.[41] When the barrage lifted off the enemy trench, the battle became a race between the attacking infantry attempting to overrun the forward enemy position and the defending infantry scrambling from their dugouts to man their defenses.[42] One memo on lessons learned on the Somme emphasized the speed at which assaulting troops had to move once the barrage lifted: "The all-important point is that the *assault must follow absolutely on the heels of the lift of barrage*. This is a matter of seconds."[43] While this was a simple idea in theory, in practice following a wall of shrapnel and high explosive shells required discipline and motivation.

Another tactical development of the fighting on the Somme was the detailing of formations as "moppers up," that were tasked with clearing outposts and trenches after the leading assault waves had passed. Given the specific function of the "moppers up," they were equipped with bomb and bayonet for the intense trench storming work. This permitted the assault waves to maintain momentum, bypassing stubborn points of resistance and continuing forward. Leap-frogging was another innovation that the Canadian Corps would introduce on the Somme in order to maintain forward momentum. This technique involved setting limited objectives for the first wave of attacking battalions, and, when these objectives were attained moving a second fresh wave of attacking Battalions through these objective to continue the advance. The Canadian Corps used this technique cautiously at Courcelette in September of 1916 with two consecutive phases of its attack. This technique would become standard practice in the Canadian Corps during 1917 and 1918 with the use of objective lines denoted by a colour (Blue, Red, Green, Black, and Brown), at which point attacking battalions would leap-frog through each other. Throughout the summer and fall of 1916, all of these refinements became common features of the fighting on the Somme.

Although costly, throughout the Somme commanders and staffs remained adaptive to changing conditions. The simplification of tactics for the first day of the Somme was a response to poor infantry training and increased shell production. The tactical changes that occurred during the Somme operations were responses to the conditions of the fighting. These new tactics were instituted quickly – in some cases within a matter of weeks. However, while these tactical changes did yield positive results, they did not guarantee success, and attacks still could, and did, result in high casualties with little or no gain.

IV. Courcelette: Tactics and Technology Old and New

The introduction of new technology continued in 1916, assimilated into the existing principles of the infantry attack. In 1914 and 1915, there was a significant proliferation of older weapon systems and techniques – trenches, grenades, rifle grenades, mortars, sapping, and even body armour in the form of the helmet, to name a few. The early years of the war also saw new roles and tactics developed for pre-existing weapons systems: the creeping artillery barrage being one example. Another example was the Lewis gun, with a portability that presented new opportunities for use of machine guns in the attack. In spite of tactical changes and the adoption of new weapons,

Lewis guns were used to achieve the same offensive goal of helping the infantry forward, so that they might assault enemy positions with the bayonet. This aim also held true with the debut of the tank at Fleurs-Courcelette on 15 September, 1916. The tank was also assimilated into the pre-war principles of the attack. By drawing fire away from the infantry, covering enemy strong points with their own fire, or by engaging in shock tactics by grinding them under their tracks, tanks assisted the infantry forward into bayonet range. That is why the first battle in the history of armoured warfare culminated with a grim hand-to-hand struggle.[44] At 6:20 a.m. on 15 September, the assaulting battalions of the 4th, 6th, and 8th Canadian Brigades advanced behind a creeping barrage lifting 100 yards every three minutes. The infantry, unfortunately, were unable to keep pace with the barrage and the Germans had time to man their front line and oppose the advance over the last hundred yards, but this resistance was incapable of stopping the attack and the German front line was assaulted. Once in the German front line, the "moppers up" went to work down the trenches and into the dugouts. J. H. Thompson, following after the assault wave with the 31st Battalion, wrote: "It was our duty in the second wave to commence on Fritz's front line and to clear it out at the point of the bayonet and bomb the dug outs."[45]

The official records of the 2nd Division contain few references to hand-to-hand combat during the first phase of the attack at Courcelette, but further investigation reveals that it did take place – and often. After losing the front line, some Germans continued to resist from shell holes. The 21st Battalion diary noted the resistance of these isolated posts: "several detached posts of the enemy were encountered and shot or bayoneted out."[46] Other sources provide further references to hand-to-hand fighting.[47] The 28th battalion history, for example, claimed that the "bayonet [was] used freely."[48] Accounts of veterans also suggest that bayonet fighting in the first phase of the attack was a more frequent occurrence than official records indicate. Sid Smith, with the 18th Battalion, recalled his first bayonet kill during the push on Candy Trench: "I'll never forget the first guy I bayoneted was a captain in the German Army. I just did it in time he had his gun in his hand and I stuck my bayonet right into his stomach, not realizing what I was doing, only protecting myself [,] see."[49] While Smith's statement suggests that the surprise and the stress of the close fight caused him to lose cognitive control over his actions, his training paid off as he instinctively delivered the killing thrust with his bayonet.

In addition to enduring the barrage, the defenders around Courcelette had been subjected to a preliminary bombardment meant to eliminate strong points and cut wire. By the time Canadian troops charged the front line trench, German soldiers had suffered under a considerable weight of shellfire. For a number of defenders, the threat of the bayonet and close combat was sufficient to collapse their morale. Contrary to the claims by the 28th Battalion history that the "bayonet was freely used," Arthur Goodmurphy, a member of the 28th, recalled: "by the time we get that close to them [hand-to-hand] their hands were in the air. They were ready to give up [,] you know [,] because they were... demoralized too."[50] Goodmurphy emphasized that the threat of the bayonet charge was pivotal in the decision of German troops to surrender.

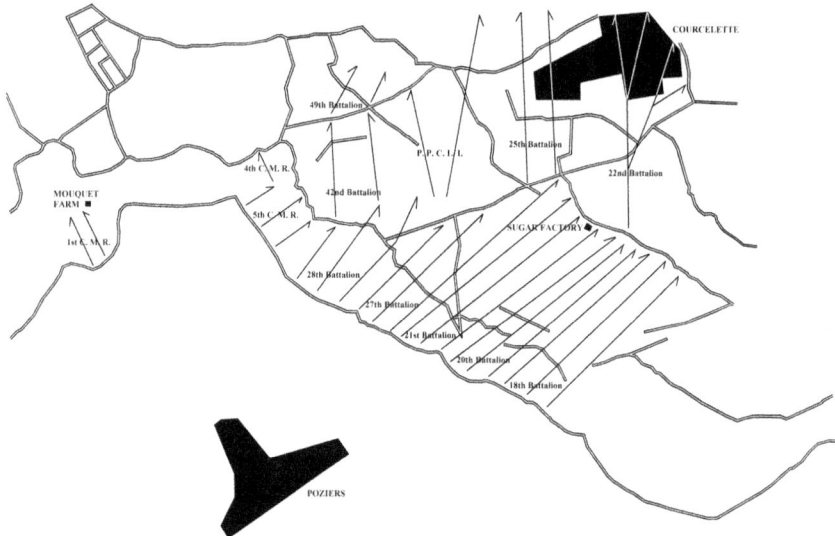

Battle of Courcelette, 15 September 1916

Artillery was not the only weapon that helped to break German morale on 15 September. Six Mark I tanks advanced with the supporting wave of the Canadian attack to assist in mopping up German positions. However, in spite of the effect the tanks had on morale – to men on both sides – their fate was typical of the unpredictable nature of armoured vehicles throughout the Great War. Of the six tanks used, one suffered from a mechanical failure before reaching the start line, one grounded itself during the course of the advance, and the other four were knocked out by German shellfire.[51] When the tanks did appear, they proved useful in assisting the infantry in their advance, clearing lanes through the barbed wire, drawing fire, and silencing troublesome strong points, sometimes by physically crushing enemy positions. But they were not a decisive factor. Only one of the six tanks reached its objective, supporting the 4th Brigade advance all the way to the Sugar Factory. Many German positions had to be cleared in hand-to-hand fighting by the infantry. Frank Maheux, with the 21st

Battalion, wrote home about the grim struggle at the Sugar Factory:

> The worst fighting here since the war started, we took all kinds of prisoners but God we lost heavy all my comrades killed or wounded, we are only a few left but all the same we gain what we were after... I went thru all the fights the same as if I was making logs. I bayoneted some [and] killed lots of Huns [sic].[52]

While the tanks succeeded in assisting the infantry forward, but the Sugar Factory still had to be secured by the men of 20th and 21st Battalions who, like Maheux, "bayoneted some" Germans.

On the left of the 2nd Division's attack, the 8th Brigade's advance against Mouquet Farm and Fabeck Graben was the scene of intense bayonet fighting. The 1st Battalion Canadian Mounted Rifles (CMR) advanced against Mouquet Farm on the left and the 5th CMR against Fabeck Graben on the right. Behind these two battalions was the 4th CMR, ready to exploit success. Advancing behind a barrage lifting 100 yards every three minutes, the 5th CMR made their objective, bayoneting 250 of the defenders and forcing a large number of Germans to flee.[53] On the right, the 1st CMR was able to get into the enemy positions and dispatch 50 or 60 Germans with the bayonet. However, the 1st CMR was later pushed out of the position and forced to dig in south of the farm.[54] The 4th CMR advanced with the second attack wave that day and managed to push further down Fabeck Graben, where it bayonetted another 150 Germans.[55] In all, roughly 450 Germans are recorded as being killed with the bayonet in the brigade report on the attack – half of the 900 Germans reported killed in the attack.[56] Reporting how Germans were killed was rare. In this instance, the bayonet figured prominently in the casualty statistics. The 8th Brigade example might also have been similar to other actions, for which no such statistics exist.

Other examples of bayonet fighting suggest that the 8th Brigade fight at Mouquet Farm and Fabeck Graben was not an anomaly. Bayonet fighting also played a significant role in the second phase of the attack as the Canadians cleared the town of Courcelette and straightened the line.[57] At 6:00 p.m., the 5th and 7th Brigades advanced behind a barrage lifting 50 yards per minute. On the left flank, the 25th Battalion advanced through the eastern half of Courcelette where it engaged in five minutes of hand-to-hand fighting.[58] The task of clearing the majority of the town fell to the 22nd Battalion, which engaged in 10 minutes of "smart bayonet fighting."[59] Captain Joseph Chabelle, with the 22nd Battalion described the experience in more detail:

> The fight at first was fierce, a hand-to-hand, no weapons barred combat:

bayonets, rifle butts, shovels, feet, teeth all came into play... Blood spurted out of punctured breasts; brains spilled out from skulls shattered by rifle butts. Bayonet fighting is a bitter struggle to the end, with no mercy shown... Oh! The sensation of driving the blade into flesh, between ribs, despite the opponents grasping efforts to deflect it. You struggle savagely, panting furiously, lips contorted in a grimace, teeth gnashing, until you feel the enemy relax his grip and topple like a log. To remove the bayonet, you have to pull it out with both hands; if it is caught in the bone, you must brace your foot on the still heaving body, and tug it with all your might.[60]

As Chabelle's account demonstrates, the soldiers view of hand-to-hand fighting was more detailed and grizzly than battalion reports and narratives usually indicate when they gloss over such events as with terms such as "hand-to-hand fighting," or the "bayonet used freely," or even "smart bayonet fighting."

New tactics and technology assisted the Canadians in achieving their objectives at Courcelette, but, in spite of these changes, the elementary principles of the infantry attack had remained unchanged. Artillery tactics had been refined with the creeping barrage to support the infantry as it closed with the enemy. The tank, when it appeared, also assisted the infantry in crossing No Man's Land. Yet, regardless of these changes, the infantry still had to charge and clear enemy positions with bomb, bullet, and bayonet.

V. The Ancre Heights: Problems Old and New

Over the next two months, the Canadian Corps attempted to batter its way through the German fortifications on the high ground north of Courcelette.[61] The infantry-artillery tactics that had succeeded at Courcelette, however, proved fragile. The rapid tempo of operations proved a constant problem in the conduct of attacks. The tight timelines for the assaults afforded little flexibility in responding to the particular conditions of each section of the front. Poor artillery ammunition and fuses hampered the tasks of wire cutting and destroying enemy positions. The increased reliance on artillery in the attack had created the additional problem of wear on artillery pieces that had continued firing long past their intended lifespan, causing a chronic problem of rounds landing either too long or short.[62] These problems also meant the infantry, now leaning hard on the barrage, frequently suffered from drop-short rounds, or from Germans unsuppressed by guns that fired long. All of these problems were further magnified by German positions being well sited on the reverse slopes of the high ground north of Fabeck

Graben and Courcelette. That clever positioning made accurate engagements of these positions difficult for all but the highest trajectory weapons in the artillery arsenal. These difficulties undermined the advantages afforded by the high volume of shells British and Canadian guns continued to throw at German positions on the Somme.

Under such conditions, the offensive problems of 1915 began to reassert themselves. Often attacks were broken up before reaching German wire, much less the German trenches. For troops who reached the enemy line, severe casualties doomed these break-ins, which were often overwhelmed by German counter-attacks.

Such was the case of the Canadian attack against Regina Trench on the reverse slope of Theipeval Ridge on the afternoon of 1 October, 1916. The reversed slope of the position created considerable difficulty for the artillery attempting to cut wire and cover the advance – shells either fell too long or too short owing to the slope. This resulted in several reports of uncut wire and additional efforts by the artillery to cut it in the hours before the attack.[63] When the attack went in at 3:15 p.m., the majority of the assaulting formations suffered heavily from a combination of German artillery, German small arms fire, and short firing Canadian guns. Most of the attacking companies were driven to ground by German fire where they either dug in or crawled back to their jumping-off positions. Some of the assaulting Companies managed to continue the advance only to be hung up on the poorly cut wire.[64] In spite of the staggering fire and wire obstacles in No Man's Land, "A" Company of the 5th CMR managed to get past the wire and assault the enemy trench:

> When "A" Coy were about 100 yards from REGINA trench enemy opened with m[achine]/gun and rifle fire which became very intense during the next 50 yards of the advance, causing many casualties. The enemy was then seen to be rapidly evacuating his front line, and very few were left when our men got up. Many of the enemy were brought down by our men as they were falling back to their rear line.[65]

As the men of the 5th CMR closed on Regina Trench and the prospect of the bayonet assault loomed, many of the defenders took to their heels only to be shot down by Canadian troops or caught in the barrage behind Regina Trench. After consolidating the trench, parties of bombers and bayonet men managed to clear 500 yards of Regina Trench to the east in order to assist the other battalions forward. In the hours that followed the initial assault, the men of the 5th CMR and the supporting 1st and 2nd CMR raced reinforcements and supplies across No Man's Land; however, the weight of German fire in No Man's Land and the constant German

Attack on Regina Trench, 1 October 1916

counter-attacks proved too much and the Canadians were ejected from the trench the next morning after almost 17 hours of "hand-to-hand fighting."[66] Lieutenant, later General, George Pearkes commanded the troops in Regina Trench and recalled in a CBC interview that Regina Trench "was the heaviest hand-to-hand fighting"[67] he encountered during the war.[68]

The 8 October attack by the 1st and 3rd Divisions against Regina Trench also encountered the persistent challenges faced by Canadian troops on the Somme. At 4:50 a.m. on the morning of 8 October, eight battalions advanced behind a creeping barrage. The Canadians managed to enter the German trenches in some places, but the Canadian artillery failed to cut the enemy wire along large parts of the front,[69] leaving several of the attacking battalions held up by the uncut wire and forced back by enemy fire before reaching Regina Trench. However, a few battalions, or at least portions of them, did manage to find their way through the wire to assault German positions. The experience of the 16th Battalion closing with the enemy trench was typical of those battalions that reached the German line:

> The situation on the right of the attack was less serious. There the artillery had smashed gaps in the wire. When the barrage lifted the companies attacking on that front rushed through into the trench. Directly in front of the extreme right flank a German machine gun was firing, but it did little damage as the aim was too high. A shower of bombs put that gun out of action; the bombers quickly behind their missiles, and were on the crew and the reserve team before they had time to defend themselves.[70]

The men of the battalion leaned on the barrage and then used a final barrage of their own bombs to rush the trench and silence enemy machine guns. After gaining a toehold in the enemy trench, the Canadians "met with fierce resistance. The enemy, the [German] Marines, fought viciously, and no quarter was asked or given by either side."[71] In the face of such stubborn resistance, the men of the 16th Battalion bombed down the trenches to gain contact with the other attacking battalions that had succeeded in reaching the enemy trenches, and they used their bayonets to carve their way through the German positions.

The RCR attack was similar to that of the 16th Battalion. The RCR kept close to the barrage, but here again the German wire proved problematic:

> Owing to the wire not being sufficiently cut, great difficulty was experienced by some groups of our men in 'A' Company and 'C' Company in getting at them, but in spite of all, they gained the desired end. Part of 'D' Company (Left Co.) however, were held up by wire and intense machine gun fire at the junction of KENORA and REGINA TRENCHES.[72]

102   *1916, The Bayonet and the Battle of Materiel*

Attack on Regina Trench, 8 October 1916

The RCR war diary gives some sense of how troops, once free of the wire, could bite into the German front line: "Then the crucial moment came, the time to clinch with the enemy. With a rush, before he had time to man his machine guns and parapet, they leapt into the trench, putting him to flight."[73] If Canadian troops could stay on the heels of the barrage, once free of the wire, they had little trouble rushing the German front line and overcoming the garrison with the bayonet. Once in the enemy trench, the men of the RCR "immediately set about their allotted tasks, bombing dug outs, collecting prisoners, and consolidating."[74] Words like "securing," "rushing," and "consolidating" obscure the bitter hand-to-hand fights that followed. H. Arnson Green, who was with the RCR at Regina Trench, recollected in a postwar interview: "Oh yes, we had physical contact, as a matter of fact I saw men bayoneted and fighting hand-to-hand in that attack on Regina Trench."[75] After wresting the trenches from the Germans, the RCR braced for the counter-attacks. As in 1915, the depleted ranks after the initial assault, as well as the slow flow of reinforcements and supplies determined the final success of an attack. After fending off three separate counter-attacks, the RCR's supply of bombs was exhausted and they were pushed out of the German front line at 9:00 a.m.[76] In desperate straits, some Canadians resorted to the one weapon they had left, the bayonet:

> Lt. Chatterton W.A. rallied these men and started a bayonet charge, but was severely wounded through the shoulder. Some of this party reached their trenches again and Lt. Chatterton W.A. who in the meantime had had his wound dressed, got up and carried on... The men were fighting with their fists. Lt. Chatterton was killed by a sniper while trying to hold the enemy back on

the right.[77]

For the men of the RCR, the ordeal of Regina Trench lasted only four hours; for other battalions the bitter fighting lasted well into the afternoon.

Such was the case of the 3rd Battalion. In the initial assault, they encountered "some opposition" but managed to take the German front line at "ten minutes after zero."[78] The battalion then bombed their way deeper into the maze of trenches and secured their second objective. For the next two hours, the 3rd Battalion consolidated and rushed bombs and ammunition forward; however, they were forced to hand much of these supplies to the 4th and 16th Battalions on their flanks, which had faced a much harder fight to enter and secure the German front line. At 7:00 a.m., the German counter-attacks began with shelling and bombing attacks down the communication trenches. Although the 3rd Battalion contested these counter-attacks, the Germans were relentless in reclaiming the lost ground. By noon, the supply of ammunition was running short and another counter-attack pushed the Canadians off their second objective shortly before 1:30 p.m.[79] On the first objective, congestion in Regina Trench from the wounded had forced the remaining men of the 3rd Battalion to defend their position from shell holes in front of the trenches.[80] They managed to hold off the counter-attacks for an hour and a half before their supplies of bombs and ammunition were completely spent. The Germans then pushed the Canadians out of the front line. Like the RCR, the officers of the 3rd Battalion attempted to counter attack with the bayonet. However, the inquest into the failure of the 3rd Battalion attack showed Canadian troops at the breaking point:

> Majors BENNETT and MOWAT... attempted a bayonet charge. A few men only went forward cheering, but were at once stopped by the enemy with bombs, machine gun fire, and sniping and the attempt failed, Major MOWAT being killed. An officer of the 3rd Battalion is stated to have shot some of our men at this point who were in shell holes and who could not be induced to counter-attack with the bayonet.[81]

This particular passage was expunged from the distributed transcript of the inquest and was crossed out in pencil on the original draft. The inquest illustrated that the bayonet charge was not just a mechanical response and spoke of the confidence required by both sides engaged in the negotiation of close combat. Attacking soldiers who lacked confidence failed to have an appreciable impact on the defenders or simply failed to launch the charge if they felt the situation was hopeless. By 3:00 p.m., the situation for the 3rd Battalion was indeed hopeless and the survivors withdrew back to their jumping off positions.

Given the friction facing artillerymen – sometimes quite literally with worn out gun barrels – the 8 October attack at Regina Trench had fallen prey to the same difficulties that had condemned attacks in 1915. From the perspective of the infantry, after actions reports laid the blame for the failure of the attack on the artillery.[82] However, even the shortcomings of the bombardment and barrage did not prevent Canadian soldiers from assaulting and clearing substantial portions of the German trenches. The unsuppressed enemy and uncut wire had taken their toll on the attacking infantry such that it lacked sufficient strength to fight off the intense German counter-attacks that followed. The bayonet charge, as in 1915, was capable – even for depleted assault waves – of wresting control of the trenches from the Germans, if only for a short time.

VI. Bayonet Training 1916: Organization and Control

In 1916, bayonet fighting continued to be an important skill, evident in the expansion of an official training regime and efforts to exert uniformity over training and techniques. Through a combination of reporting, increased control over instructors, and even through disciplinary measures, the chain of command was able to gain control over the chaotic state of bayonet training that had developed in an *ad hoc* fashion.

Through the efforts of Major Henry George Mayes, the foundations of the administration had been laid at the end of 1915. Mayes had begun the organization of Canadian bayonet fighting schools in Canada and at Canadian training camps in England. In February 1916, a Canadian bayonet fighting school was established at Shorncliffe training camp. The school was given an establishment of: " Lt. Col. Director and Inspector, 2 Capt. Superintendants., 4 Lt.,2 Army Staff instructors in Bayonet Fighting and P.T., 14 Sergeant Major Instructors in Bayonet Fighting and P.T., 2 Orderly Room clerks," and "4 Batmen."[83] Mayes's efforts had attracted the attention of the chain of command and he was promoted to Lt. Colonel and appointed Director and Inspector for Physical Training and Bayonet Fighting in February.[84] The establishment of Bayonet fighting schools in Canada also continued, and, by April 1916 additional schools had been established in: London, Calgary, Toronto, and Halifax.[85]

Soldiers had received 20 hours of bayonet training until the first half of 1915, when it was reduced to nine hours. In November 1915, this was deemed insufficient and increased again to 24 hours over the course of a 10-week syllabus. In October 1916, it was increased again to 60 hours training over a lengthened 14-week syllabus.[86] The increased training was not limited to new recruits; once their basic training was completed, soldiers outside of France had to engage in half-hour training sessions in bayonet fighting six days a week.[87] Even for troops in France, training in bayonet fighting was a regular part of training syllabi while battalions were in reserve. By 1916, official training systems had begun to meet the front-line demand for bayonet fighting.*

Much of Mayes's work to increase bayonet training relied upon a concomitant increase in battalion bayonet instructors. Assistant instructors did have other regular responsibilities within their battalions, but the increased staff – even if only part-time – facilitated bayonet training. In February 1916, sixteen platoon bayonet instructors joined the four company instructors in each battalion. In addition, two officers in each battalion were given the responsibility of overseeing bayonet instruction, bringing the total strength of a battalion's bayonet training staff to 22.[88] Within the Canadian Corps, this increased establishment was achieved by the end of May 1916.[89] Also all instructors now had to attend a 21-day instructor's course run by the CAGS at Shorncliffe, or by the AGS at Aldershot.[90] This system gave the newly-established Directorate of Bayonet Fighting and Physical Training, under Mayes, direct input in the training of the assistant instructors in each battalion.

These changes created a strong cadre, or community, of bayonet instructors within the battalions of the CEF. Not only were the cadres trained, they were the eyes and ears of the Directorate. Monitoring was the key to control and uniformity over bayonet training. This gave the instructors a support network to fall back on and permitted unofficial techniques to be observed and corrected directly within the battalions. Not only could these battalion instructors police themselves, but through the filing of status reports and weekly, or sometimes daily, bayonet returns – which reported training activities in the battalion – the Directorate was able to police the instructors as well. Through this system of regular reports the Directorate took measures to eliminate undesirable techniques and practices. For example, on 20 February 1916 the Army Council issued orders to stamp out the informal practice of twisting the bayonet by disciplining the bayonet instructors, and making them financially responsible for rifles and bayonets broken by improper use in training.[91] Through control and surveillance, a uniform regime for bayonet training was established.

The Directorate also published two pamphlets for bayonet training

---

* However it seems the Canadian units were using the 14 week syllabus and an increase number of hours devoted to bayonet training before October 1916. For example, the 85th Battalion training file includes a 14 week syllabus with a total of 39 hours of bayonet training dated July 3rd 1916.

during 1916. The first was *Bayonet Fighting 1916*, which introduced several new techniques for bayonet fighting and instituted the "blob stick" or "parrying stick" as an official training tool. *The Organization of Bayonet Fighting and Physical Training in a Battalion C.E.F.: 1916,* was the second publication. It clearly defined the equipment, organization, and training of men for bayonet fighting. This work also included a series of lectures meant to inculcate the "spirit of the bayonet" and emphasized the importance of physical training. The Directorate also published a series of established syllabi to assist instructors in the execution of training.[92]

Figure 9: The blobstick in use (Bayonet Fighting 1916)

The blobstick had been an informal training tool since 1915. Now it became an official fixture of training. The official blobstick was a 1¼ to 3 inch wide stick that was between five and six feet long with a two or three inch metal ring on one end and a cloth or leather bag (the blob) filled with rags at the other. Some blobsticks included a point that extended beyond the ring. Parries were made against the stick, points were thrust into the ring, and butt strokes executed against the padded blob. This provided a manipulated target that did not risk damaging the bayonet or rifle.

The blobstick also reintroduced the partner, or opponent, into bayonet training, thereby making up, at least in part, for many of the training shortcomings since the loss of "loose play" in 1914. The Blobstick gave soldiers the means to practice defensive techniques with the bayonet and rifle against a moving opponent. Nonetheless, the Blobstick was not a perfect substitute for "loose play" and bayonet fencing. Private Thomas Dinesen lamented how bayonet fencing, as part of training would have helped:

> If they would only put each of us into a padded coat and fix a wooden stick on to the rifle instead of the bayonet, we should be able to fence and fight to our hearts content and learn all sorts of useful tricks; whereas now we simply thrust into empty air, stolidly and indifferently.[93]

The blobstick for all of its merits could not give the complete experience that men received by actually fencing with spring bayonets. According to the 1916 establishment, each battalion possessed six full sets of bayonet fencing gear, hardly enough for training, but enough for at least some of the men to spar.[94] George Hedley Kempling wrote home about his positive experience bayonet fencing in the summer of 1916. "We did bayonet fighting with a heavy dummy rifle and a spring bayonet, and wearing a padded suit and a big helmet. We went at it hammer and tongs and played as dirty as we could for we couldn't hurt the other fellow very easily..."[95]

In terms of techniques, *Bayonet Fighting 1916* added the *feint* to the soldier's arsenal. By using a false attack, the bayonet could be circled underneath the opponent's parry and the *long point* delivered. The idea of *feint* in bayonet combat existed in nineteenth century bayonet fencing, but this was the first time the technique appeared in wartime training literature. The *feint* also reintroduced the opponent's actions in bayonet fighting into training. The reintegration of the opponent was in part due to the introduction of the blobstick, but also to bayonet fencing, which had continued as a limited sporting activity.

The development of *infighting*, which began with the 1905 system of bayonet fighting, evolved further in 1916. *Bayonet Fighting 1916* introduced two new techniques of *infighting* the first was attacking with the trigger guard or magazine and the second was the use of the knee. The first involved "Smash[ing] the magazine or trigger guard violently into the opponent's face."[96] This was a refinement of the butt stroke, to be used if the opponent was inside the distance of the butt. The knee was to be used "When gripped by an opponent and unable to use the point." The technique involved "the knee [being] brought up against the fork [of the legs] or the heel stamped on the instep..."[97] These techniques added to the arsenal of the bayonet fighter.

The establishment of controlled and uniform training system was not simply the enforcement of a static system. The official sanction in 1916 of techniques and tools – the knee and the blobstick – that were unofficial in 1915 demonstrated that the Directorate was meant to be the conduit for the assessment of front line techniques. They also reviewed the unofficial technique of twisting the bayonet, and made efforts to stamp out the practice. With an effective training system in place by the summer of 1916, the Directorate then ensured the uniform application of these techniques across the Canadian Corps. The establishment and operation of the Directorate demonstrated not only the growing importance of the bayonet, but also the importance of fostering and controlling front-line innovation. With the training establishment organized, bayonet fighting would adapt to the changes heralded by the developments of the set piece battles of 1917.

Chapter VI: 1917, The Bayonet and the Set-Piece Battle

THE UTILIZATION OF new technology and significant shifts in tactical thinking had an effect on the battlefield of 1917. In spite of these changes, there were also considerable points of continuity between the fighting on the Somme and Canadian set-piece battles of 1917.[1] New weapons introduced in the first years of the war permitted the refinement of infantry fire and movement. New technology also permitted artillery counter-battery operations to be fully integrated into offensive operations. In spite of significant tactical changes in the set-piece battle of 1917, the bayonet continued to play a significant part in operations. Bayonet fighting conformed to tactical changes within the Anglo-Canadian forces which emphasized increased independent action on the part of individual soldiers, low-level initiative in dealing with unforeseen situations, and integration with the expanding arsenal of the infantry platoon.

I. Vimy Ridge: Change and Continuity

There were a number of important tactical changes for the infantry after the Somme, foremost of which was the improved application of fire and movement tactics. This went hand-in-hand with the devolution of control that *dispersion* tactics required. The 16th Battalion history laid out the tactical developments between December 1916 and April of 1917:

> In the attack units, instead of moving into battle on a broad front in a succession of straight lines as previously, went forward on a narrow front; companies and battalions attack one behind the other in an irregular formation which gave them the maximum protection from artillery and machine gun fire and enabled them to take full advantage of ground and weapons. When the first objective was captured, the second wave 'leap frogged' the leading troops and pressed on to the enemy's second line of defence; and, that captured, the succeeding waves passed through to further objectives. The plan was an elaboration of the Somme method of attack, with the new principle that movement conformed to a pre-arranged timetable instead of light signals... thereupon all hastened to give these [platoon] officers and their men the range of weapons and organizations that best suited to their important work. One, and the two Lewis guns, the automatic rifles which had come into use, rifle, bayonet, rifle grenade and bomb were all placed at the disposal of the platoon; the Lewis gun – a weapon of opportunity – and the rifle to deal with the enemy in the open, the rifle grenade and bomb to get those behind cover, and the bayonet for hand-to-hand fighting. With the combination of these weapons, each supporting the advance as the need arose, it was possible for the commander and his men to initiate tactics suitable for a variety of conditions and ground. In other words, the purpose behind this grouping was to create a balanced, self-sufficient fighting body which could act as the spear head of the attack, ready at a moments notice to exploit the advantage of battle.[2]

Canadian operations in 1917 used the platoon as a capable independent sub-unit in battle, especially for "stalking" strong points. Over the course of that year, a renewed emphasis on musketry training would re-establish the rifle as a useful tool in these devolved fire and movement tactics. Yet, in spite all of these developments the principles of the infantry attack still conformed largely to *Infantry Training 1914*.

The combination of fire and movement, by its nature, relied on the initiative and independence of subalterns and NCOs who were responsible for organizing and executing these small-scale attacks as they arose. Between the Somme and Vimy Ridge, Brigadier-General Percy Radcliff, of the Canadian General Staff and Lieutenant-General Arthur Currie acknowledged this importance of low-level initiative:[3]

> It may be pointed out that there is nothing new in this system of training. Before the war we endeavoured to make the platoon a self-sufficient unit of battle. Owing to the demands for so many specialists, there grew up in our battalions a wrong system of organization and the development of Company, Platoon, Section and Squad leaders were some-what neglected. It is necessary

for us to revise our own training* on the old lines.⁴

The preparations for the Battle of Vimy Ridge acknowledged this relationship and fostered initiative. The designation of platoon objectives and careful rehearsal of attacks gave junior leaders the confidence to rely on their own initiative and continue moving forward.⁵ As Currie observed, this was not a new idea. The use of fire by one group of infantry to assist others in moving forward had been advocated in British training literature since 1902 and known as "fire and movement."⁶ It had been a part of Canadian training as early as August 1914.⁷ The actual techniques used by the infantry in closing with the enemy remained unchanged, and the infantry of 1917 advanced in short bounds or rushes as had been practice in British training since 1905. Prewar *dispersion* thinking had predicted the requirement of decentralized control in the execution of fire and movement, and, as the level at which the groups skirmished decreased, so too did centralized direction of the movement.⁸

There was also a noticeable shift in artillery tactics from destruction to neutralization. The purpose of the neutralizing artillery fire was to win the firefight with the enemy and keep him under fire rather than destroying German fortifications, which could never be fully accomplished.⁹ The refinement of the creeping barrage during 1917 continued to allow the infantry to lean on the barrage and then rush enemy positions as soon as the barrage lifted. Ideally, the distance between the barrage and the assaulting infantry was fifty yards or less – ranges suited to the use of bomb and

---

* This is also found in a memo from Lt. Gen R. Butler: LAC RG 9 III-C-3, vol 4031 Folder 26, file 7 "Training: 1st Canadian Infantry Brigade," Correspondence G.B./165, dated May 10th, 1917:

> An impression appears to have arisen in some formations that the instructions issued from time to time by G.H.Q. France, (such as S.S. 135 and 143) either supercede, or are at variance with the principles contained in the manuals.
>
> This is not the case, and it is the duty of commanders to see that the principles laid down in the manuals are adhered to. It must be clearly understood that the pre-war manuals remain in force and that the instructions issued by G.H.Q. are merely amplifications of these manuals in order to meet the varying requirements of this campaign.
>
> For instance, one of the objects of the new platoon organization and training was to reduce the undue importance which was being given to so-called specialists, I.e. the use of the rifle as opposed to the use of bombs, and it was pointed out in the G.H.Q. instructions referred to above that the rifle and the bayonet were the first weapons of every infantry soldier and that all ranks must be proficient in their use.
>
> Again, the principles of the employment of fire to cover movement remains unaltered, whether the covering fire is given by artillery, rifles, machine guns, rifle bombs, or Stokes mortars. In this connection recent experience indicates that the employment of covering fire from rifles and machine guns, when distant objectives are being attacked, requires more attention.

bayonet. The foundations of artillery in the attack had changed, but the intention remained to support infantry closing on enemy positions.

New technology also provided artillerymen additional opportunities to help ensure the success of the attacking infantry. The development of sound-ranging and flash-spotting permitted the artillery to conduct counter-battery operations. Counter-battery operations specifically targeted enemy guns, thereby reducing enemy shellfire during the attack. It also reduced the enemy artillery available to support German counter-attacks against newly-won positions.[10] Counter-battery fire too was harnessed to aid the infantry as they closed with the enemy.

All of these changes were present at Vimy Ridge. The preparation and planning of the infantry for the attack relied on independent action, initiative, and devolved control. Objectives had been set for individual platoons, using the platoon as the "self-sufficient [sub-]unit of battle" identified by Currie. Platoons were prepared to fight their way forward independently toward objectives, and this placed considerable responsibility for the conduct of the attack in the hands of subalterns. Frequent rehearsals for the attack over taped courses provided soldiers with confidence in their knowledge of objectives. This confidence, in turn, permitted the infantry to improvise in the face of unforeseen circumstances without losing sight of their objectives or awaiting orders from higher levels. Rehearsals also permitted platoons to continue forward as subalterns became casualties, and NCOs took charge of the attack. Everyone knew what to do.

The artillery preparation for the assault on Vimy Ridge involved counter-battery fire and the continued refinement of the creeping barrage. In the weeks preceding the attack, Canadian Corps Counter-Battery Office used flash spotting and sound ranging to plot German artillery positions and permitted the accurate engagement of German guns and mortars before the attack. In addition, the fire of hundreds of guns were plotted "minute by minute"[11] for a creeping barrage that lasted several hours and, in some cases, included several dozen lifts. The artillery preparation succeeded. Counter-battery fire silenced 83% of German artillery on 9 April, and the creeping barrage assisted Canadian infantry in achieving the majority of their objectives on schedule.[12] The destructive preparatory bombardment, a fixture of the 1915 and 1916 battles, remained a prominent part of the artillery preparation for Vimy. The artillery conducted a heavy week-long preparatory bombardment, meant to shatter German defenses and break the morale of the defending troops. The German description of the preparatory bombardment as the "week of suffering" was a testament to the weight of Canadian shellfire and its value in cracking the defender's morale.[13] The demoralizing effect of the standing barrage was also demonstrated by the four thousand Germans who were taken prisoner at Vimy, this represented roughly 20% of the total number of prisoners captured by the Canadian Corps before August 1918.[14]

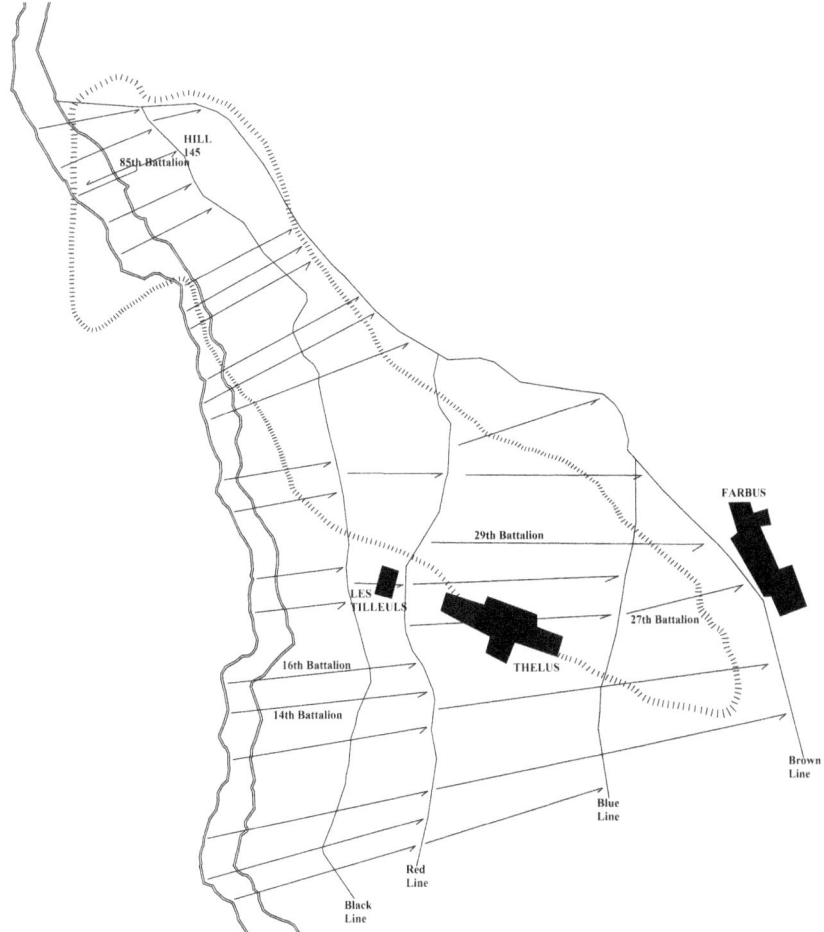

The Battle of Vimy Ridge, 9 April 1917

As successful as the artillery was, it remained a tough battle for the infantry, who still had to close with the enemy. The artillery had succeeded in destroying much of the German defenses, but pockets of defenders held their ground and much of the initial resistance came from scattered strong points that had survived the bombardment. The result was described by the 16th Battalion history as "a running fight, men rushing from shell-hole to shell-hole, the bodies of the fallen indicating by their position the locations of the enemy's gun towards which this fighting was directed."[15] On the right flank of the 16th Battalion, the 14th Battalion also encountered German resistance that had survived the bombardment: "Many of the enemy

(Bavarians) fought strongly to the last, showing no inclination to surrender."[16] Grenades were used to silence two of the machine guns. The third was silenced by Lieut. E. F. Davidson armed with a revolver, after dodging German grenades.[17] The last gun "on the left was charged by Company Sergeant Major J. F. Hurley, who unassisted, bayoneted the crew of three men and captured the gun."[18] For the men of the 14th Battalion, the barrage may have permitted the infantry to advance on the enemy positions, but four machine guns still had to be silenced at ranges of 30 yards or less.

Lewis guns and bombs helped the infantry get close enough for a final bayonet charge against these scattered points of resistance.[19] At the Brown line, one of the Canadian objectives, the men of the 27th and 29th Battalions used rifle grenades to suppress German machine gun fire and create the opportunity for a bayonet charge:

> The machine guns were dealt with by the rifle grenadiers most successfully, and our troops of both Battalions, raising a loud cheer, charged for the final 50 yards, leaped down in to the German gunners in their gun pits and trenches. A stout fight was put up by these Germans, urged on by several Officers, but all were soon bayoneted or captured by our troops and their guns were in our hands.[20]

The ability of the infantry to advance independently using fire and movement was demonstrated in the hasty attack of the 85th Battalion against Hill 145. After the failure of the first wave formations to reach their objectives, the 85th Battalion was ordered to attack Hill 145 at 6:45 p.m.; however, the barrage was cancelled just before zero hour and the 85th was forced to attack unassisted. Under its own strength, the 85th fought its way forward: "The attack was pressed home, the Companies providing their own suppressing fire by Lewis guns firing from the hip and riflemen firing on the move."[21] Bounding forward, using fire and movement, the Canadians succeeded in closing on enemy positions and then charging home with the bayonet. The battalion history described the actions of Captain Percival W. Anderson in the hand-to-hand fray: "[Anderson] captured several machine guns and was always in the open inspiring his men by his dauntless courage. He engaged himself in hand-to-hand combat encounters with the enemy fighting with pistol and bayonet and sometimes with his fists."[22]

The Canadian attack on Vimy Ridge differed from the fighting at the Somme in many regards. The complexity of the creeping barrage, the accuracy of predicted fire, and the development of counter-battery fire all represented a considerable refinement in the skills of gunners. The infantry had also further assimilated Lewis guns and rifle grenade in the attack, giving the infantry platoon the tools it needed to conduct small-scale infantry attacks against points of resistance. In spite of these changes there remained aspects of the Canadian attack at Vimy Ridge that conformed to the tactics of 1916. The crushing weight of the preparatory bombardment cracked fortifications and undermined morale, but the infantry still had to close on enemy positions and silence them.

II. The Arras Offensive: Change Continues

After Vimy Ridge, the Canadians pursued the Germans into the Douai plain. Here, the Corps was forced to adapt to changes imposed by the new system of German defence. At the beginning of 1917, the Germans had embraced greater *dispersion* in their defensive tactics; however, the narrow heights at Vimy Ridge had not permitted the defences facing the Canadian attack to conform to the new system. After Vimy Ridge, the Canadian operations at Arleux, Fresnoy, and Souchez faced the new German defence in depth.[23]

Trench lines and dugouts still dominated the German fortifications, but they had diminished in importance during battle. The new defensive system permitted lower troop densities in the forward zone of operations, reducing casualties under the crushing weight of Entente shellfire.[24] The larger-level organization of German defences established large defensive areas, sometimes several thousand yards in depth.[25] Each of these areas served different functions. The forward area was defended by strong points, rather than continuously held lines; these were intended to disorganize attacking formations before they reached the second, or main, defensive area.[26] The second defensive area was held in greater force and meant to stall attacks, already broken up by the resistance of strong points in the forward area. Behind these two areas was a third area, manned by formations tasked with counter-attacking Entente penetrations into the forward and main defensive areas.[27] The heart of the 'elastic' system of defence was the counter-attack.[28] This increased *dispersion* and depth involved an increased devolution of control, and local commanders were given greater latitude in how they defended against attacks or organized their own local counter-attacks.[29] This left the Entente forces facing large-scale *dispersion* as the Germans deployed over wider defensive areas, often out of range of effective artillery interdiction. Entente troops had to fight their way through the dispersed forward area before reaching the main line of resistance, after which they faced counter-attacks by German troops that had not been subjected to bombardment. The British suffered greatly from these new German defensive tactics when the second phase of the Arras offensive opened on 23 April.[30]

Concurrent with changes in German defensive tactics, the Canadian Corps also imposed changes on its offensive tactics. The fighting of 1916 and 1917 had revealed that soldiers had relied too heavily on the grenade.[31] Arthur Currie commented on bombing after the Somme:

> Bombing: Our training in this regard has long been on absolutely correct lines. While I want every man in the Battalion to be a bomber, I want them to realize that there are *other* ways of killing a Bosche than by exploding a bomb underneath his feet. It is in actual fact that so accustomed have some of our men been to bombing that they have chased a Bosche hundreds of yards in order to get close enough to him to hit him with a bomb whereas a bullet would have done the trick just as well.[32]

Nor was it merely observed by generals, infantrymen also observed this trend. A veteran of the 49th Battalion recalled in a CBC interview for the radio series *In Flanders Fields*:

> They started to train us... to use our rifles again... You had to keep it clean but you never used it... At one time it got so that the bomb was more important that the rifle you see. They had to get the men back on to, to use their rifle and improve their musketry.[33]

The over-reliance on the hand grenade had led to a lack of confidence in the rifle.[34] Even before the Arras offensive had ground to a halt, both British and Canadian command instituted a new training system to increase the confidence of troops in their rifles, as described in this memo issued by the British First Army:[35]

> As our operations approach more and more the standard of open fighting to which *we were trained before the present war*, it becomes more and more important that the men should know how to use their rifles and that junior officers and N.C.O.s should be quick to seize the opportunities of using rifle fire especially in stopping small local counter-attacks on the flanks of the attack.[36]

Renewed emphasis on the rifle, platoon self-sufficiency, and the initiative of subalterns and NCOs increased steadily over the next year-and-a-half. The increased emphasis on musketry training quickly took hold in the Canadian Corps and the importance of rifle fire in the assault, and in the defense, increased in Canadian operations throughout the second half of 1917.

III. Hill 70: Charge and Counter-Charge

The refinements in Canadian attack doctrine were demonstrated at the attack at Hill 70 on August 15. This operation combined the counter-battery techniques developed in the opening months of 1917 with the meticulous planning of Vimy. On the tactical level, the attack used devolved platoon tactics and broke up repeated German counter-attacks with artillery concentrations, along with infantry relying on rifle and Lewis gun fire. In spite of the new tactics, Hill 70 was the scene of much bayonet fighting and close combat.[37] For example, of the 42 Military Medals awarded to members of the 31st Battalion in the fighting at Hill 70, half were for actions involving hand-to-hand combat or for "rushing" enemy positions.[38]

At the beginning of July, the Canadian Corps was ordered to attack the city of Lens on 30 July to divert German reserves from the opening phase of the Ypres offensive, set to begin the next day. The onset of heavy rains, however, postponed the operation by more than two weeks.[39] Arthur Currie, who had been given command of the Canadian Corps at the beginning of June, adhered more to the spirit of the order than the letter. Eschewing the option of a grinding street fight through the city of Lens, Currie succeeded in convincing British command to have the objective changed to the high ground of Hill 70, just north of the city. This limited objective attack was to break into the German forward defensive area, secure and consolidate the Hill 70, the high ground overlooking Lens, and then grind down the inevitable German counter-attacks. It was to be a battle of body counts and not ground taken.

The preparation and initial assault made full use of the tactical refinements of 1917. In the two weeks preceding the operation, Canadian artillery softened up the enemy defenses, cut wire, and neutralized 40 percent of the German artillery within range of Canadian guns.[40] At 4:25 a.m. on 15 August, the creeping barrage commenced and 10 battalions of the 1st and 2nd Divisions attacked. The leading battalions advanced through the German front line to the first objective at the Blue Line. At 5:20 a.m. five additional battalions pushed forward across the Blue Line in the center of the attack, deeper into German positions to the Red Line. Finally, in the center, the 10th and 5th Battalions, which were in the leading wave, again leap-frogged through the Red Line to capture the Chalk Quarry at the Green Line. That is how the battle looked from the Corps level.

At the lower levels there was a lot of close and intense fighting. The advance to the Blue Line was characterized by small groups of Canadian infantry bounding forward and rushing dispersed strong points. A detailed description of one such action appeared in the 13th Battalion history:

118 *1917, The Bayonet and the Set-Piece Battle*

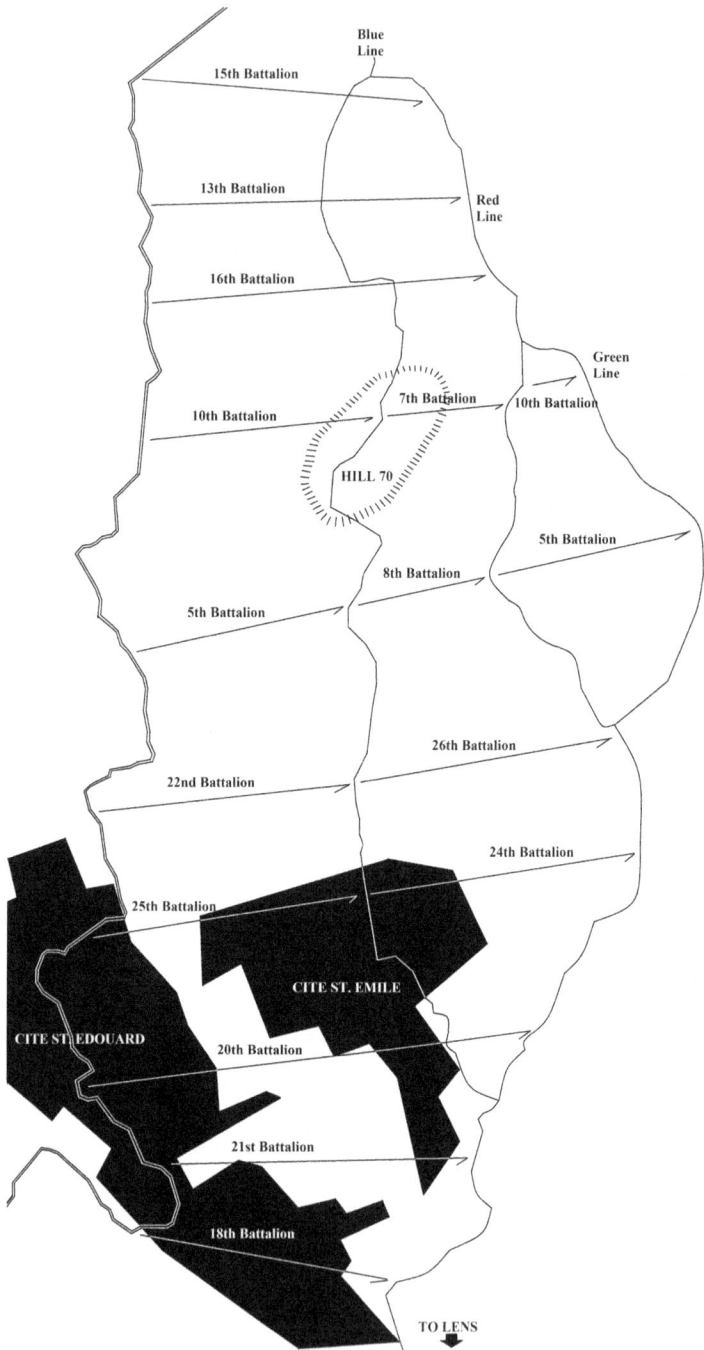

The Battle of Hill 70, 15 August 1917

With his diminished numbers, (now Major) Mathewson decided that an attempt to rush the position would be dangerous. Accordingly he ordered the men of 'D' Coy. to stalk the enemy, creeping ever nearer and nearer, firing all the time and closing in until a rush would seem advisable. Although these tactics involved a slight loss of time, they succeeded admirably. Noticing that a few of the enemy had started to retreat, Mathewson decided that the psychological moment had arrived and gave the order to charge. Responding with a yell, the Highlanders plunged forward and took the trench with the bayonet, killing or wounding many of the garrison and capturing around a dozen of unwounded prisoners.[41]

Here, the bounding approach of the Canadian infantry was sufficient to make the Germans consider the consequences of what followed. The official account of this bayonet charge also highlights the ambiguity of the contemporary records on the subject of bayonet fighting. The battalion diary blandly recorded: "There was no fighting in the Front Line System, or BLUE LINE and the number of prisoners was small, about 25 prisoners."[42] The important demoralizing effect of the bayonet often went unrecorded – in the War Diaries at least.

In these small engagements, Canadian infantry used their firepower to suppress enemy positions while sections or individuals bounded their way forward to a point from which hand grenades could be thrown and followed up with the bayonet. The 7th Battalion War Diarist recorded "We soon came under machine gun and rifle fire, principally from a machine gun which had to be silenced by rifle grenade and rifle fire. The position was eventually rushed and two or three prisoners were sent back from here."[43] The increased emphasis on musketry since May had taken hold. "The rifle appears to have been made more use of than at any time since bombs first made their appearance,"[44] wrote one 4th Battalion officer. Within three months, the new training regimens had begun to have an appreciable effect on front line operations.

In the storming of the German strong points and fortifications, the pace at which close combat occurred forced soldiers to quickly improvise with whatever resources were at hand. One instance, in the 15th Battalion, was the use of a clubbed Lewis gun:

The Corporal [Robson], Lewis gun on his shoulder, charged alone, straight for the woods and disappeared. The same berserk N.C.O. was prominent at the Blue Line. Suddenly he noticed he had forgotten his revolver and in the midst of the chaos he began to swear. As he leapt into a trench a German officer bounced out of a dugout doorway with his nice blue-steeled Luger pointed at the Corporal's head. Here was a heaven sent opportunity. The Corporal clouted the Hun over the head with his Lewis gun barrel, took the revolver and gaily proceeded on his way.⁴⁵

Nor was the enemy soldier the only target of the shock effect and close combat. The 10th Battalion report recorded a hand-to-hand attack on a machine gun:

> During the advance, Cpl. P[E]RMAN of the 10th Battalion, saw an enemy machine gun being put into action. He charged the gun and was shot through the right arm, but before the gun could be got into action he seized the barrel and tipped it over backwards down the sap from which it had been brought. An Officer and 6 O.R. then surrendered to him.⁴⁶

Here, the threat of fighting at close quarters was enough to break the enemy's will to resist, even if the immediate threat was a single wounded infantryman. In close quarters fighting, the bayonet and the rifle butt were not the only weapons in the soldier's arsenal. The training literature published over the summer and fall of 1917 reflected this chaotic nature of close combat and began training soldiers to use weapons other than the bayonet, rifle, and legs in individual close combat.

At times Canadian troops used bayonets to hold on to newly-won positions. At Hill 70, the German counter-attacks began on the exposed left flank of the newly won Canadian positions at 12:45 p.m. on 15 August. Grinding down these counter-attacks was the key Canadian goal. On the north end of the Canadian line, the 13th and 15th Battalions faced many of the counter-attacks launched by the Germans over three days. They quickly depleted their supply of small arms ammunition as they fended off repeated thrusts. At one point, ammunition had become so scarce that the men of the 13th Battalion made use of a bayonet counter-charge to break up a German attack that had made it through the defensive barrage:

> Major Macfarlane found to his anxiety that his rifle ammunition was down to 10 rounds per man, even after every possible cartridge had been collected from casualties. Finding that the officer in command of the company of the 16th Battalion on the right flank was facing a similar shortage, he agreed with the latter that no further firing should take place. A few minutes later an attack developed up Humbug Alley, a communication trench connecting the Green Line with Hercules Trench. In accordance with the agreement, no rifle fire was used to check this attack, No. 14 platoon going over to meet the Hun with the bayonet and one section, filled with more zeal than discretion, pursuing the defeated enemy right back to Hercules Trench.⁴⁷

The counter charge, by bayonet, broke up the German attack. *Assault Training 1917*, issued a month after Hill 70, also advocated the counter charge under just such conditions:

> In order to prevent being taken at a disadvantage, the holders of a position without ammunition, must make a controlled and well-timed counter assault and push it home with all the vigour, dash, and determination of fresh men against an enemy who has been subjected to an advance under trying circumstances... The one [side] stimulated with the greater fury and confidence, by the force of its determination to conquer, will cause the other line to waver and turn.[48]

This reiterated the concept of *élan*, the idea that the side with the greatest offensive spirit would prevail.

Similar use of the bayonet in the defense was demonstrated all along the Canadian line. Another example occurred south of Hill 70 in the early morning hours of 18 August, when the Germans attacked the front held by the 20th and 21st Battalions. The 20th Battalion diary described this fight:

> He got into the trenches held on our right, and into COMMOTION Trench between the LENS-LaBASSEE road and NABOB ALLEY... parties were collected from 'A' and 'B' Coy's and the enemy engaged with rifle and bayonet. Not content with putting him out of COMMOTION and part of CONDUCTOR Trenches, the men from these two companies followed him out into No-Man's-Land in hand-to-hand conflict.[49]

The fighting lasted for over an hour and the situation was reported "normal" at 5:40 a.m.[50]

The bayonet proved useful in desperate defensive circumstances. During the initial German counter-attack, Private G. A. Fuller was part of a Lewis gun post at a trench block in Nabob Alley, when a shell landed killing all but Fuller, who was buried with the Lewis gun. Sergeant Frederick Hobson, who was nearby when the shell hit, dug out Fuller and recovered the Lewis gun. Fuller and Hobson fired the Lewis gun at the advancing Germans until the weapon jammed. Hobson, now wounded, left Fuller to clear the stoppage and went forward to hold back the Germans closing on the trench block. Hobson then "with bayonet and clubbed rifle, single handed[ly] held them back until he... was killed by a rifle shot."[51] Hobson's single-handed bayonet charge gained enough time for Fuller to get the Lewis gun operational, after which he continued to hold off the Germans. When the attack was finally beaten back, Hobson's body was found amongst 15 enemy dead. Hobson was awarded a posthumous Victoria Cross for his action.[52]

In the defense, the bayonet also relied on the forward movement with which it had been advocated in British training since 1905. This was a direct application of *élan* by attempting to break up an attack by applying a greater offensive force. This could be used after an enemy's offensive momentum had stalled, as was seen is the counter-attacks against Germans who had entered the Canadian trenches. Another defensive use of the bayonet was the counter-charge. The counter-charge relied on a defensive bayonet charge into the preparation of an enemy bayonet charge; here each side would attempt to demonstrate a greater offensive spirit in order to break the morale of their opponent. In the case of Sergeant Hobson, the use of the counter-charge could also sap the forward momentum of the enemy, permitting the organization of defenses or additional counter-attacks.[53] In essence, the use of the bayonet in the defense was to attack in order to recover the offensive initiative or momentum from the enemy, and these techniques were amply demonstrated in the fighting at Hill 70.

One of the more notable bayonet fights at Hill 70 occurred on the morning of 21 August, in the fighting around Nun's Alley. Here, according to official records, "the value of the bayonet and of bayonet fighting training was once again strongly emphasized."[54] The flurry of German counter-attacks had wound down by the evening of the 18 August; however, in spite of their failure to recapture Hill 70 and an estimated 20,000 casualties, the Germans refused to relinquish Lens. The Canadian Corps attacked Lens itself on 21 August in the hope that the German will to resist had been broken in the previous week's fighting. That hope proved to be a misplaced.

The attack was ordered for 4:35 a.m. on the morning of 21 August. The 10th Brigade was to attack the city from the west as the 6th Brigade pushed into the city from the north. The 6th Brigade objectives were Combat Trench, Cinnebar Trench, and Nun's alley. At 3:00 a.m., as the 6th Brigade formed up for the attack, the Germans began a bombardment that increased in intensity until 4:30 a.m., when the Germans launched an attack of their own. Moving behind their barrage, they assaulted the front line held by the 25th Battalion (on the left) and the 29th Battalion (on the right). On the left, the German infantry, having less distance to cover, fell upon the 25th Battalion's front line just before the Canadian zero hour, forcing the men of the 25th Battalion from their positions. On the right, at 4:35 a.m., the 29th Battalion advanced from under the German shelling they had endured for more than an hour to follow their own creeping barrage. Almost immediately, 'D' company of the 29th Battalion came under heavy machine gun fire from the active German positions in Nun's Alley. At this point, the two attacks – Canadian and German – collided in No Man's Land, and fierce hand-to-hand fighting occurred between the barrages.[55] On the left of the 29th Battalion attack, 'D' company inflicted heavy casualties on the enemy, but, between the machine gun fire and the bayonet fighting, the company was obliterated. 'C' Company, on the right, also inflicted heavy losses on the Germans during the 15-minute bayonet fight and succeeded in pushing the Germans back into Nun's Alley.[56]

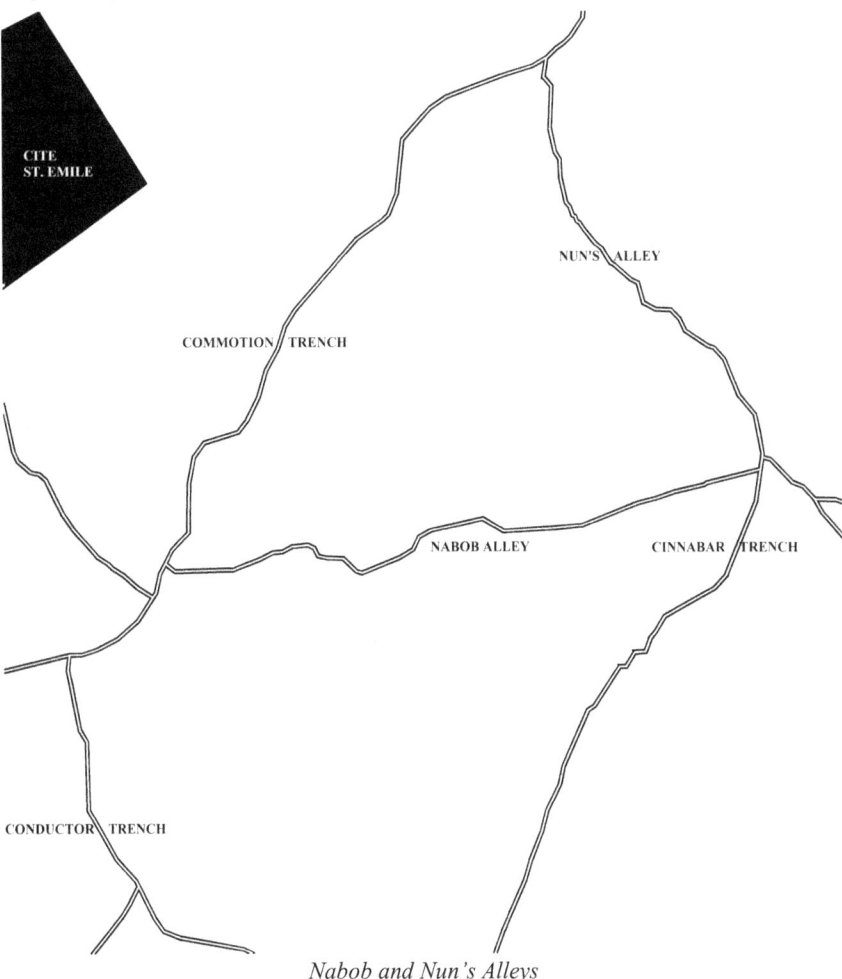

Nabob and Nun's Alleys

Once in the German front lines, the hand-to-hand fighting continued as the two sides engaged in bombing duels in which the bayonet was a critical weapon. Company Sergeant Major Robert Hanna recalled the fighting in Nun's Alley that day:

> I run into these Hinney's there, one big fellow and coming along of course all I had to do was pull the trigger, but then another fellow was coming so fast he ran into the bayonet. I didn't know if after that fellow came, the other was coming there behind the Hinney, so I got him and I thought, oh god there goes

my... my bayonet you know, I hit him in the head...[57]

Hanna's account demonstrates the stress, confusion, and unpredictability of hand-to-hand fighting. Hanna encountered the Germans moving down the trench, who themselves seemed initially unaware the Canadians had entered the trench. Hanna also had to contend with his bayonet becoming dislodged and was forced to use his rifle butt against the last German. Hanna's somewhat modest and vague recollection belies his bravery and courage that day, for which he was awarded the Victoria Cross.

The record of Hill 70 provides ample evidence of Canadian soldiers using the bayonet in the defense and the continued importance of maintaining offensive momentum in counter-attacks. Additionally, Hill 70 demonstrated the refinements in offensive tactics developed on the Somme: the creeping barrage had succeeded in neutralizing and suppressing German fire, counter-battery fire had eased the weight of shellfire faced by attacking infantry in the assault and consolidation phases of the attack, and the renewed emphasis on musketry had permitted the rifle to play a more significant part in both the attack and the defence. All of these adjustments assisted the infantry in closing with the enemy to assault with bomb and bayonet.

IV. Passchendaele: Individual and Independence

Throughout 1917, the Germans increasingly based their defense on dispersed concrete pillboxes in the forward defence area to break up assaults.[58] Three years of constant fighting in the Ypres Salient had also destroyed the water drainage of the region and the heavy rains of 1917 had turned much of the area into a mire of mud and water-filled craters. Artillery pieces sank into the mud and shells drove deep into the soft ground before detonating, lessening their effect. For the infantry, the conditions reduced movement to a slow pace and assaulting battalions suffered heavily as they inched forward through the mud under German fire. In some cases, the mud clogged rifles and jammed Lewis guns making the advance using fire and movement even more difficult.

Dispersed groups of infantry assaulting strong points and pillboxes were not unique to the Canadian experience of the 3rd Battle of Ypres. A number of reports from the British experience of fighting at Ypres were issued to the Canadian Corps for the purposes of training and preparation. A British First Army report on lessons learned from the 3rd Battle of Ypres, illustrated German reliance on pill-boxes and the fire and movement tactics used to overcome them:

During the recent fighting the garrison of a concrete block-house situated some 150 yards in advance of a certain stretch of captured ground were seriously interfering with our front positions by keeping up a constant and accurate fire upon anyone showing himself; they succeeded in killing and wounding several of our men, and the Company Commander decided that it was necessary to make an attempt to capture the block house. A Corporal in the company, realizing the situation, *on his own initiative* posted some men to give him covering fire by shooting at anyone moving near the block-house, whilst with 4 other men he himself worked round a flank. The covering party fired steadily, whilst the Corporal and his men worked unobserved from shell-hole to shell-hole until they reached a position well on the flank of the block-house. They then made a rush for it and completely surprised the garrison, killing 6 Germans and capturing 18 others, including an officer.[59]

This particular small action followed broadly the principles of the attack set down in *Infantry Training 1914*, with one group of men using rifle fire to force the enemy behind cover so that another group of infantry could close by bounds, until, finally, the pillbox was rushed with the bayonet.

Throughout the Canadian phases of the fighting at Passchendaele there were similar recorded instances of fire and movement tactics in the stalking and rushing of pillboxes, strong points, and dispersed groups of German infantry.[60] Gordon Brown, with the 46th Battalion on the right flank of the Canadian attack of 26 October, recalled one such action against an isolated machine gun:

Finally... one of the boys discovered the bullets were coming from down in the swamp to the very left of the advance. We had by-passed him on our way up in the morning. He was well hidden – almost impossible to get a shot at him with a rifle – and one shot a time was all we could get out of our rifles without cleaning the bolt by pouring water or urinating on them. There was a young fellow with me by the name of Donald Dickin from Manor, Saskatchewan. He had poor eyesight and wore thick-lensed glasses. He was in the rifle-grenade section and said, 'Show me the blighter, Brownie, and I'll get him.' When I pointed out where this gunner was, Dickin said he was out of range; he'd have to get closer. So us fellows covered him the best we could with our rifles while he made a dash to get within range of the machine gunner. When Dickin got into a shell hole I took after him and joined him, and again pointed out where the enemy gunner was. Dickin thought he could get him with a long-range rifle grenade, so he got what he thought was the right angle on it and pressed the trigger.[61]

The Battle of Passchendaele, 26 October to 6 November 1917

Rifle fire, limited by the mud had permitted only Dickin and Brown to close individually on the enemy machine gun. Once in the effective range of the rifle grenade, Brown used it to cover his next movement forward:

> I watched the grenade go up in the air, turn, and come straight down beside the enemy gunner. As soon as the grenade exploded I was on the run to get him before he could get his wits about him. But I need not have been in a hurry. Dickin was as good as his word – he had got him.[62]

These coordinated actions – with rifles and Lewis guns supporting the movement of the rifle grenadiers, who in turn supported the advance to bombing range – permitting the final rush with the bayonet. Here, the rifle was used to keep enemy heads down while Dickin bounded forward with the rifle-grenade, which he then used to neutralize the enemy before

Brown's final rush. But, the position still had to be secured by closing with the bayonet.[63]

Close-in fights were difficult and often ruled by chance and luck. On the left flank of the 26 October attack, the 4th CMR engaged in small level infantry assaults on individual pillboxes in a fashion similar to that described in the First Army report from September:

> The first line of pill-boxes was reached on time with the barrage. Owing to heavy going the barrage appeared to be properly timed. Heavy opposition was met at the line of pill-boxes and heavy casualties, particularly amongst officers, were experienced in rushing these strong points. The enemy fought bravely, keeping his rifles and machine guns in action until bayoneted or bombed.[64]

Lieutenant Tom Rutherford was one of the officers involved in rushing the first line of pillboxes with the 4th CMR, and his account provides an illustration of the chaotic hand-to-hand fighting that occurred around these fortifications as small groups of soldiers worked their way forward:

> The barrage had just lifted. We saw no Germans around nor were we shot at. It was one of those pillboxes with a shallow trench close behind it running out to both left and right, with a wooden lean-to covering the trench to keep out the rain. The Sergeant – Sergeant Nicholas of 'C' Company – had his bomb ready in his hand and I motioned to him to go in first. We both jumped into the shallow trench and underneath the third board cover we came to an opening, where he threw the bomb in. Almost instantly, and before the bomb had gone off, a big wild-eyed German without weapons jumped out between us and, in his fright, turned on me under the low roof; he grabbed me by the head and started twisiting it. I had my revolver in his chest but he didn't let go and I was beginning to black out. Having seen other Germans coming out and following Nicholas out the other side I pulled on him and in falling on top of me, he pushed me back to the opening.[65]

As Rutherford attempted to extricate himself from the grasp of the dying German soldier more Canadian soldiers appeared in the melee and amply demonstrated the element of random chance that reigned in close combat situations:

> So here was a sweater-clad figure backing out from the lean-to to the pillbox, with a dead German partly on top of him. Just then an excited man arriving with some others from 'C' company, was loading his rifle so he brought the butt down on my head, but as the German had pushed the rim of my steel hat up on edge I was none the worse.[66]

The tension of advancing through the enclosed spaces and sharp corners of trenches and pillboxes, soldiers attacked out of reflex when a possible threat appeared.

The 85th Battalion attack on Vienna Cottage in the early hours of 30 October demonstrated the importance of fire and movement tactics. For the 85th, the barrage overshot the target and the battalion was forced to work their way forward with ineffective artillery support. They were instantly met with rifle and machine gun fire from dispersed German positions, and a series of small independent firefights developed. Undeterred, the men of the 85th provided "their own covering fire with rifle grenades, Lewis guns and rifle fire..."[67] and continued to advance. According to the battalion history, small groups of men disposed of individual machine gun nests, often by rushing these positions with the bayonet:

> When our advance was held up on the left of Decline Copse by machine gun fire and heavy artillery action[,] Lance-Corporal Kenneth P. Harris led an attack on an enemy machine gun post with rifle grenades, and when near enough a few well-placed Mills bombs finished the Hun resistance and the bayonet did the rest.[68]

Here rifle grenades were used to cover the infantry's advance, and Mills bombs provided the final neutralization of enemy fire for the bayonet charge:

> Private Alexander McDonald was another man who took charge of his section after the section's commander had become a casualty, and by means of bombs and rifle grenades succeeded in penetrating a nest of enemy machine guns and personally attacked and bayoneted three Huns who were resisting.[69]

These were bayonet charges led by NCOs. Few things could have demonstrated better the depth to which the devolution of initiative and control had penetrated the Canadian Corps.

The success of these individual firefights for strong points and machine gun nests helped tip the balance for the 85th Battalion, as the war diary recorded: "the line was drawn closer and closer to the enemy by the men jumping from shell hole to shell hole."[70] As the Canadians approached, by bounds and rushes, the German resistance slipped past the tipping point. Now came the moment for the bayonet charge: "As the resistance slackened our men with a loud cheer rushed upon them with the bayonet and soon dispatched those who had not escaped in precipitate flight."[71] But this too was a bayonet charge organized and led by NCOs and small groups of men:

There were no other officers there, and Corporal John H. Campbell was the next senior N.C.O. He immediately took command of the platoon and led it on with great dash until checked by a nest of enemy machine guns. With coolness and daring he quickly organized his platoon for a charge and led them in the face of a desperate fire right up to the guns, capturing them and destroying the enemy.[72]

The men of the 85th Battalion demonstrated all the traits of prewar infantry tactics that began to be practiced after the Somme: the use of inherent firepower to cover, the advance in bounds by small groups of soldiers, and the devolved control by subalterns and NCOs. However, all of these tactics were used to close to the range at which the bayonet charge could be launched.

The official record of the attack, written from the battalion's perspective, acknowledged the combination of Lewis gun fire, rifle fire, and rifle grenades covering the infantry advance, but the records failed to appreciate the significance of the bayonet:

> Whether the enemy saw the reinforcements arriving, or whether the rifle grenades, Lewis guns and sniping had overpowered the enemy is difficult to say; but it appears that the enemy broke just as 'D' Company, with Major Anderson, came up; and the whole line swarmed across the hostile front line...[73]

The battalion report described the individual instances of bayonet fighting and the final charge ambiguously, as 'swarming' over the enemy position.

Nor was this the only bayonet fighting by the 85th Battalion that morning hidden by the language of official records. After clearing these initial German positions, the battalion continued towards its final objective. At Vienna Cottage, they encountered 15 to 20 defenders with 2 light machine guns supported by a field gun. According to the official records, they "rapidly cleared"[74] the objective, but once again, this clearing was executed by small groups and individuals charging home with the bayonet:

> Private Waslin E. Kirkirikos... pushed forward in advance of the severely threatened right flank, and seeing an enemy field gun in operation some distance ahead made for it. There were two Germans in different shell holes armed with automatic rifles with large magazines, who immediately fired on him. He bayoneted the first man and shot the second and was the first to reach the field gun and captured it single-handed.[75]

The official records frequently glossed over the individual acts of bayonet fighting, most likely because the authors of these official reports took for granted that those reading these records knew that a successful infantry attack concluded with the bayonet charge. The men of the 85th Battalion had demonstrated all the recent adjustments to infantry tactics: the use of platoon weapons to suppress the enemy and permit the advance in bounds, as well as the devolved control to subalterns and NCOs. All of these tactics were used with the aim of closing to the range at which the bayonet charge could be launched.[76]

V. Bayonet Training 1917: Bayonet Fighting Matures

While the bayonet assault remained a constant principle of the infantry attack, the techniques and training of soldiers for close combat was undergoing considerable refinement in response to the increased emphasis on devolution of control and independent initiative during 1917. Three new training pamphlets were issued during the summer and fall of 1917 reflecting these changes.

The first of these manuals was *Methods of Unarmed Attack and Defense* published in June 1917. Not strictly for infantrymen, this manual was meant for wide-scale training of soldiers. In addition to combat techniques, it included sections on the lifting and carrying of wounded soldiers and a section for military policemen illustrating how to subdue unruly "friendlies." The techniques outlined in *Methods of Unarmed Attack and Defense* were a departure from previous systems of close combat. This work focused on situations arising if a soldier had lost his bayonet. This pamphlet taught soldiers how to use their hands, feet, and even steel helmets to subdue opponents attacking them with fists, knives, or bayonets. These were techniques required by individual soldiers who managed to get amongst the enemy and prepared them to continue fighting in a much wider variety of situations.

The second pamphlet was a Canadian publication by H. G. Mayes entitled *Bayonet Fighting Illustrated 1917,* which appeared in August of 1917.[77] This manual was meant to complement *Bayonet Fighting 1916* and contained illustrations for a number of techniques covered in previous pamphlets. For instance, several pictorial examples of the proper use of the blob stick were provided, as well as numerous examples of the techniques of using the rifle butt for bayonet fighting.[78] In addition, *Bayonet Fighting Illustrated 1917* added a new set of techniques that relied on the actions of an opponent. These likely began as some of the "dirty tricks" developed from the introduction of the blobstick and the limited bayonet fencing that still occurred within the Canadian Corps. The *duck*, for example, involved making a false attack and then lowering the entire body to deliver the point underneath the opponent's weapon.[79] *Bayonet Fighting Illustrated 1917* also added new techniques to the growing arsenal of *infighting*. Some, such as a number of techniques for unarmed grappling and the disarming of a bayonet-wielding opponent, were repeated from *Methods of Unarmed Attack and Defense*, but new *corps et corps* techniques were also included. One taught soldiers how to apply the *jab* while in physical contact with their opponent.[80] Another showed soldiers how to bind their enemy's bayonet in order to close within *infighting* range and then use a puttee knife to stab the opponent in the groin.[81] Emphasis on *infighting* went hand-in-hand with the tactical change away from the bayonet charge in waves to the individual soldier driving forward on the dispersed battlefield.

While the first two bayonet publications of 1917 emphasized the role of the individual in hand-to-hand fighting, the third emphasized the cooperative use of the bomb and the rifle to bring the soldier to bayonet range. Published in September 1917, *Assault Training 1917* listed the bayonet and the rifle as the chief weapons of the infantryman: "The bullet and the bayonet belong to the same parent, the rifle, which is still the deciding factor on the battlefield. One must work with the other."[82] *Assault Training 1917* did not add any new techniques to the arsenal of bayonet fighting -- however, it did provide a series of exercises meant to train with the combination of the rifle, bomb, and bayonet on a single course. The first exercise was an elaborate course that presented the soldier with a wide variety of targets for practice:

> The soldier leaves the starting line as the barrage lifts, and walks at a steady pace to the charging line. Firing from the hip may be practiced during this stage if considered desirable.
>
> On reaching the charging line a whistle is blown and he breaks into a double. Then he sees a sniper in hiding in a shell hole, bayonets him and proceeds to get through the [wire] obstacle.
>
> This accomplished, he is met by a German represented by a dummy whom he bayonets. Whilst he is engaged with this dummy, another German climbs on the parapet and is about to shoot him. The soldier turns immediately and puts a bullet through the target which is held up by an operator in the trench at this moment. He then charges the trench with a shout and bayonets three dummies in the trench.
>
> As soon as he gets into the trench, he finds he is being bombed from a shell hole beyond, locates the shell hole and throws two bombs into it. He then recharges his magazine and, finding he cannot shoot over the parados, he overcomes the difficulty by making a firestep. As soon as this is complete he then fires five rounds rapid at the target on the bank.

> If the man be a rifle bomber, he can be made to carry his cup attachment in his bomb carrier until he arrives in the objective trench when he should be made to employ rifle bombs against a suitable target.[83]

Soldiers were put through this course at the base training schools, like Le Havre, on their arrival in France.[84] This training course permitted the soldier to train with a combination of weapons as they practiced the tasks of assaulting, consolidating, and defending against the counter-attacks. This exercise course also demonstrated the growing emphasis on musketry training in the last half of 1917 by training in the use of the rifle in both the assault and the defence. As well, this exercise acknowledged changes in German defensive tactics and forced the trainee to dispatch German defenders dispersed in shell holes in front of, and behind, the trench.

VII. Conclusion

The small-scale application of fire and movement and bounding by small groups of infantry in clearing German strong points indicated the depth to which initiative and independence had penetrated the Canadian Corps by the end of 1917. Yet in spite of the impact of these important concepts of *dispersion* tactics (devolution of control and independence of action), the basic principles of the infantry attack remained unchanged. Fire was used to cover the infantry closing in bounds or rushes on enemy positions, which were then subjected to the bayonet assault, forcing the defenders to fight, flee, or surrender.

Far from being a static traditional weapon, the bayonet proved adaptable to new tactical conditions. The most significant evolution in infantry tactics during 1917 focused on devolved control in offensive operations, such as the concept of the "self-sufficient" platoon, and the emphasis on fire and movement. The chain-of-command acknowledged the importance of the platoon and section as self-sufficient sub-units of battle, led by subalterns and NCO's. The devolution of the battlefield in 1917 was also evident in the changing techniques for bayonet fighting and the refinement of *infighting*. This new impetus for preparing the individual to fight in close combat and in a wide variety of situations mirrored refinements in infantry tactics over the course of 1917.

A by-product of platoon self-sufficiency was an emphasis on the cooperative effect of infantry weapons, with the bayonet remaining an essential one of many. Arthur Currie acknowledged the place of the bayonet in the infantry platoon arsenal: "Confidence was born of good training. The men knew how to shoot straight and how to use their bayonets... They knew how to use the bomb, the rifle Grenade and the Machine Gun, but best of all they knew the effective combination of these weapons."[85] The Canadian Corps in the final offensives of 1918 put this "effective combination" repeatedly to the test.

Chapter VII: 1918, The Bayonet and the War of Movement

AFTER THE FIGHTING at Passchendaele wound down, the Canadian Corps shifted to manning the line for almost nine months and would not engage in any major operations until August of 1918. In this time the experience of the Canadian Corps also diverged significantly from the British Imperial experience of the winter, spring, and summer of 1918. In the early months of 1918, due to manpower shortages British Imperial Forces reduced the size of an infantry brigade from four battalions to three, and, the strength of an Imperial infantry division dropped from 12 battalions to nine.[1] However, the Canadian Corps chose to maintain the strength of Canadian divisions by breaking up the Canadian 5th Division – which was training in England – to provide replacements.[2]

The Canadian Corps was also spared the onslaught of the German Spring Offensives which began on the 21st of March 1918. While Canadian cavalry and the Motor Machine Gun Brigade were active in operations to stem the German attack, General Currie resisted attempts by GHQ to commit Canadian infantry divisions to counter attacking the German penetrations in the British line.[3] Currie was grudgingly forced to release the Canadian 2nd Division from the Canadian Corps. The 2nd Canadian Division spent three months holding defensive positions in the British line permitting British units to be released for offensive actions elsewhere.[4] However, the Canadian Corps was by no means inactive and the bayonet saw use in the frequent trench raids and combat patrols the conducted by Canadian troops between December 1917 and August of 1918.[5]

The Canadian Corps also used the period between November 1917 and August 1918 to train for "Open Warfare." The changes in the training syllabus in 1918 emphasized physical endurance and continued independent initiative of platoon and section commanders that came to prominence in 1917.[6] George Pearkes, who became a Major-General in the Second World War, commented on the training for open warfare of 1918:

> I think it was depending more on the initiative of the junior leaders and it was a case of where you had got out of the trenches and made a break through and then, instead of merely advancing a few yards which was all we'd ever been doing in practically linear formation, then we were going to carry on the advance into miles in which you might be more or less in continuous contact with the enemy although, towards the end, you might have lost contact with them and so you were training to develop the leadership qualities in the platoon's section commanders, company platoon section commanders. Instead of having everything cut and dried for them.[7]

The change from trench warfare to the pursuit to Mons required significant training. Given finite training time, priorities had to be shifted in order to emphasize the aspects important to "open warfare" and some elements of training had to assume a lower priority. There was an increase in Musketry training and as a result bayonet fighting assumed a lower priority in training in 1918; however, bombing assumed an even lower priority.[8]

In comparison to the British Imperial units that had seen their bayonet strength reduced and worn down containing the German Spring Offensives, Canadian Division were fresh and had a significant edge in combat strength over their British counterparts. These factors made the Canadian Corps well suited to play a key role in the Imperial offensive at Amiens after the Germans returned again to the defensive in the summer of 1918.

I. Amiens: The Bayonet and Open Warfare

After the German offensive had stalled Entente forces planned a series of offensives to force the Germans from their newly won positions before they could be heavily fortified. The French were the first to strike at the 2nd Battle of the Marne at the end of July.[9] The British attack fell in early August near Amiens on the German gains that had retaken the ground that had been hard won in 1916 in the fighting on the Somme. The Canadian and Australians spearheaded an attack featured new tactics developed and tested in the British attack at Cambrai in late November of 1917. The Imperial troops would be supported by a massive deployment of tanks, more than 600. Amiens featured only the creeping barrage, relying on surprise and the tanks to carry the Canadian advance forward.[10]

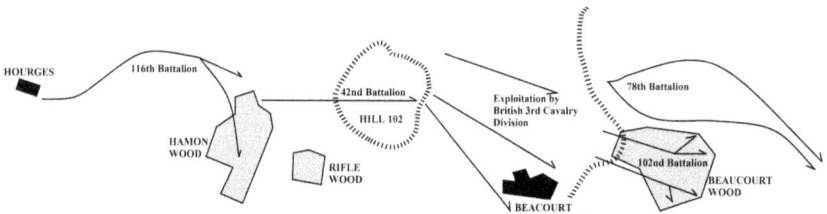

The advance of the 116th, 42nd, 102nd, and 78th Battalions during Amiens, 8 Aug. 1918

The Battle of Amiens featured several dozen Canadian hand-to-hand fights that have been identified from multiple sources.[11] Examining only a single portion of the advance will demonstrate the continued important of the bayonet in Canadian attack doctrine – in particular the attacks by 116th, 42nd, 102nd, and 78th Battalions. The attack began at 4:20 a.m. on 8 August and, behind a creeping barrage, the 116th Battalion drove forward with bayonets fixed into the forward German defenses. The battalion history hinted at the hand-to-hand fighting that occurred as the men of the 116th "rushed"[12] several machine gun nests and cleared the forward German defensive area and Hamon Wood. F. G. Thompson, a bayonet instructor for the 78th Battalion, advanced behind the 116th that morning and witnessed the bayoneted Germans left in the wake of advancing Canadians:

> And this was a sight to me because I was B.F.P.T. [bayonet fighting and physical training] for the 78th... But to see a bunch of Henies piled up in a conventional attitude correctly described that was bayoneted by the first wave, oh those boys must have had a great time at that.[13]

For the 116th Battalion, this hand-to-hand fighting was described in the brigade narrative as: "at many places opposition developed, but was quickly overcome."[14] The use of euphemistic language to document bayonet fighting in official records was nothing new.

At 8:20 a.m., the 42nd Battalion leap-frogged through the 116th Battalion and up the slopes of Hill 102. A battery of German 4.1-inch howitzers firing over open sights held up the left and center of the advance. To overcome this position, the men of the 42nd enveloped the battery. The war diarist wrote: "two platoons of the centre Company worked their way round the right flank until they got in rear of the battery when they opened up with Lewis guns at 100 yards range."[15] The 42nd Battalion history described this action in more detail:

> Meanwhile 'B' Company in the centre and 'C' Company on the left pressed on with equal success. They were held up for a time by a battery of 4.1 Howitzers firing point blank into the attacking platoons, but this opposition was rapidly overcome by a daring movement carried out by Captain J. D. MacLeod... took charge of two platoons of 'B' Company... and succeeded in working around the right flank to a point in the rear of the battery from which they opened fire with Lewis guns at a range of one hundred yards. The gun crews still gallantly fighting three of their guns and making frantic efforts to get the fourth one out were thrown into confusion. The attacking platoon closed in with a rush among the retreating gun teams, one of which was already hitched up. Most of the enemy immediately surrendered...[16]

In order for prisoners and guns to be taken, the work of the Lewis guns was followed by a rush. In this case, 'B' company had used fire from Lewis guns to suppress the enemy as the infantry closed to bayonet range. The 42nd Battalion war diary continued: "A little later, on the right flank, 'A' Company were held up by a battery of 8" Guns firing at point blank range from the valley... Working up within a short distance the guns were rushed and the crews were either killed or taken prisoners."[17] The use of the bayonet is implied by the use of the word "rushed," and is confirmed by the battalion history:

> The platoons went forward in rushes straight towards a battery of eight-inch Howitzers, which the enemy gallantly kept in action until the last moment. Reaching a point about one hundred and fifty yards short of the battery position, Captain Trout sprang up and led his wildly shouting company down the slope in a headlong bayonet charge which completely over-ran the battery, the crew of which was scattered in confusion and all were killed or captured.[18]

Here, the Canadians were confident enough to dispense with fire and movement and charge 150 yards. Soon after, 'B' and 'A' company had each secured an enemy battery, 'C' company rushed another strongpoint with the aid of three tanks:[19] "A platoon of 'C' Company on the left, carried out a similar operation against a high velocity gun which was holding up the advance from a position in Peronne Wood, [they] out-flanked the position, rushed it and killed or captured the whole of the crew."[20] Bayonet charges against artillery batteries were executed a number of times throughout the Hundred Days.[21]

An attack by the 102nd Battalion on Beaucourt Wood demonstrated the fire-and-movement techniques being used in conjunction with the new technology of the tank. The battalion leap-frogged through the 42nd Battalion at 12:10 p.m. and was tasked with clearing Beaucourt Wood, which was east of Hill 102. As the battalion closed on its final objective, it met stiff German resistance:

> On approaching BEAUCOURT WOOD these two companies came under terrific Machine Gun and Trench Mortar fire and owing to the fact that the two Tanks allotted to us had not yet caught up with our rapid advance they were held up for some little time. Then two Tanks from the 54th Bn. front came to our assistance and our men were enabled to continue their advance, but this could only be done in section rushes under covering fire from supporting sections. When within fifty yards of the wood they charged and captured the edge by storm, taking numerous prisoners and Machine guns.[22]

The tanks helped the infantry advance to bayonet range: "[the] two whippet Tanks gave us great assistance, enabling us to engage the enemy hand-to-hand..."[23] This account of the fighting at Beaucourt Wood reiterated the unpredictable nature of tanks on the battlefields of 1918, as the tanks that assisted in this attack had actually been allotted to another battalion, and the two tasked with supporting the 102nd Battalion had failed to appear. While mechanical reliability had improved since the debut of the tank in 1916, German anti-tank tactics succeeded in disabling a large number of the tanks during the first day at Amiens. As a result, tanks frequently failed to support the infantry. When tanks did appear, however, they managed to help the infantry forward.[24] Even at the end of August 1918, Canadian command reiterated "that their [tanks] action must be auxiliary to that of the infantry."[25] Canadian infantry had been given some training in cooperation with tanks, but it had been emphasized that the infantry were not to rely on tanks in order to achieve their objectives.[26] The "self-sufficient" infantry platoon still had to depend primarily on its own techniques of bounding and fire-and-movement to close with the enemy.[27]

At roughly the same time as the 102nd Battalion launched its bayonet charge on Beaucourt Wood, the 78th Battalion was moving forward in bounds. The 78th also leap-frogged the 42nd Battalion at 12:10 p.m., on the left flank of the 102nd Battalion. The battalion advanced through the valley to the east of Hill 102 and up the slope beyond. Here, they were confronted by heavy fire from a German strong point on their right. The after-action report gave a brief account of the clearing action that ensued:

> In order to permit of the main body of the Battalion continuing on its objective, I ordered 'C' company under Lieut. Tate to attack in front of the strip of wood to the east of BEAUCOURT en SANTERRE thus forming a right flank. This operation was successfully carried out and resulting in many casualties to the

enemy and upwards of twenty prisoners being sent back. Twelve machine guns counted here.[28]

The battalion diary attested to the threat this position presented to the right flank of the advancing companies – even a single machine gun on a battalions flank could stop an attack in its tracks. But the bayonet charge and hand-to-hand fighting that led to the awarding of Edward Tate's Victoria Cross was left out of the record. Oscar Ericson was one of the men who participated in this "successfully carried out" attack:

> [Tate's] company came up against an obstacle. A German machine gun nest was hidden in a clump of woods that faced his particular company and they had to stop and they had to get right down to the ground and keep there under cover, or more or less, or as close to the ground as they could until finally the colonel came up. And our colonel was the type of man who would not brook any kind of interference with an ultimate aim, which was to get forward and capture a position. So when he came up to Tate's company he spoke to Tate and said 'What in the hell's holding you up?' So Tate said 'well there's a machine gun out there'. So the Colonel said 'Well I'm going to order you to go on and take it.' And Tate said to the colonel 'Do you realize what this will mean that practically there will not be a man to survive this?' And the colonel said 'I don't care you've got to clean it up because you're holding up the entire Battalion.' So Tate said 'O.K.' he gave his instructions that they were to go forward in shallow rushes, five men at a time and each one was to go forward at his own particular time when he thought it opportune, rush forward, get down to the ground again and another section would get up and rush forward and get down to the ground again and finally they got up within a matter of yards of this particular machine gun nest. They were blazing away all the time and finally Tate decided that this was the time and he gave the blood-curdling yell 'Come on 78th' and got to his feet beckoned his men on and all the men that were still able to go got to their feet and rushed in with him, their bayonets at the point and that's how they captured this particular position, which was heavily armed with machine guns and that's how Tate won his Victoria Cross.[29]

Ericson's observation that "each one was to go forward at his own particular time when he thought it opportune" was a clear indication of the importance of initiative and independence in fire and movement tactics. This incident also hints at the lethality of close combat and the intimidation of an attacking troops closing on its objective. The twelve machine guns captured suggests the position was manned by 70 to 100 Germans, yet, only 20 were taken prisoner, the rest had fled or were killed. This kind of stalking of machine gun nests and strong points and then charging with the bayonet was not an exceptional occurrence; it was a staple of the Amiens fighting.[30]

II. Scarpe: The Anatomy of a Bayonet Fight

The attack by the 7th and 8th Brigades in the opening phase of the Arras offensive on 26 August provides one of the most detailed Canadian records of hand-to-hand combat as the Princess Patricia's Canadian Light Infantry (PPCLI) assaulted the German positions west of Monchy-le-Preux. John William Lynch, an American serving with the unit, left a detailed account of a bayonet fight in his memoirs of the war.

The objective of the PPCLI was a fortified line of the main German defensive area; however, in accordance with German defensive doctrine, these trenches were deep in the German position, nearly 5,000 yards east of the Canadian front line. Between the Canadian line and the main German line, lay the scattered German outposts meant to sap the momentum of the Canadian attack and impose disorder on assaulting formations.

The initial phase of the attack began at 3:00 a.m. on 26 August spearheaded by the 8th Brigade assault on the German forward defensive area centered on Orange Hill. The 5th CMR created a diversion by attacking the hill frontally, while the 4th CMR pushed into the German lines north of the hill to protect the flank of the main attack. The 2nd CMR, under the cover of darkness, enveloped the hill and swept over the defenders from the north flank. The attack was successful, and the German strong points were silenced one by one with "bomb and bayonet."[31] The 1st CMR continued the advance and clear the village of Monchy-le-Preux by 7:00 a.m. in the second phase of the attack.

At 9:45 a.m. the 7th Brigade then pushed through Monchy-le-Preux to continue the third and final phase of the attack against the German main line of resistance. The 7th Brigade attack was to have four tanks in support; however, all four failed to reach the start line. The two allotted to the RCR, in the center, were knocked out by enemy fire, and the two supporting PPCLI, on the left, suffered mechanical failures.[32] In addition to the failure of tank support, the RCR had considerable trouble with the tall grass:

> Moving off the eastern edge of Monchy-le-Preux, through long grass, scrub, and a maze of old trenches and barbed wire, the three front line companies of the Regiment encountered heavy enemy shellfire and intense machine gun fire from hidden positions on their front and from strong points on both flanks. Despite this and the resulting losses, they drove determinedly forward, mopping up a number of machine gun posts hidden in the long grass and

fighting hand-to-hand at the many points where enemy infantry was encountered.[33]

In spite of clearing several of the hidden strong points, the RCR attack stalled shortly after 11:00 a.m. in the face of heavy machine gun fire.[34] For the rest of the day, the long grass continued to plague the RCR, as German counter-attacks used it as cover to infiltrate forward and rush RCR positions from close range.[35]

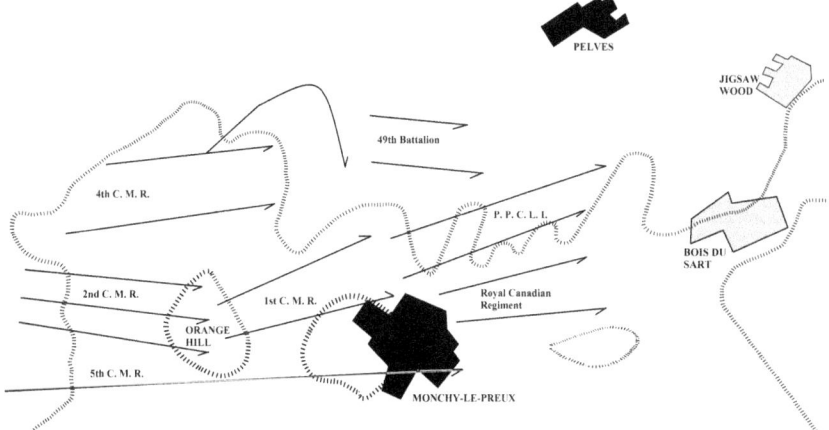

The advance of the 7th and 8th Brigades during the Battle of Scarpe, 26 August 1918

The PPCLI succeeded in reaching and assaulting the main German lines. There they met with "some opposition"[36] in Milan Trench before pushing into Faction Trench. At that point, machine gun fire from the area of Fuel Trench (the PPCLI's final objective), the village of Pelves and the area of Jigsaw Wood stalled the attack around noon.[37] Once again, the ambiguous language of the official records obscured the nature of hand-to-hand fighting. John William Lynch's account of the assault on Milan Trench elaborated on the "opposition":

> I was nearing the first trench [Milan], near the bottom of the hill. Halfway up the hill was another trench [Faction] and along the summit still another [Fuel Trench]. Our men were just beginning to rush the first trench.
> When rushing a trench the last several yards must be made in one charge. If a man tries to work his way too near a trench going from shell hole to shell hole, he stands a good chance of being blown to pieces by a bomb.[38]

Lynch then made his final approach on the German trench. Unfortunately, he was spotted by a German machine gun that had deployed behind the trench line, and he was forced to ground in front of the parapet.[39]

The machine gun then switched to fire on other small groups Canadians advancing in rushes. Given a respite from German fire, Lynch regained his composure and entered the trench:

> Gripping my rifle tightly in my right hand I half rolled, half leaped into the trench. Before I fully cleared the parapet I saw two Germans in the bay, one on the firing step, the other in the trench. Both were half turned from me. Without bringing my rifle to my shoulder I fired at the man on the firing bench who was whirling toward me. My bullet passed through his chest from right to left, killing him instantly.[40]

Lynch had disposed of one German by firing from the hip, as advocated by *Assault Training 1917*,[41] but he now had to contend with the second. Lynch became sensitive to the increased pace of being at close grips with the enemy: "No time to reload. I drove at the second man with my bayonet. He turned barely in time to parry my lunge and I crashed into him from the force of my rush."[42] This use of forward motion had been reinforced in British bayonet training since 1905.

Lynch's speed, combined with the German turning toward him, closed the distance between the two so quickly that Lynch did not have the opportunity to deploy *shorten arms* or the *jab* and the engagement passed directly to *infighting*:

> He dropped his rifle and threw his left arm around me. Almost before I realized what he was doing his right hand fumbled at his belt. Out came the sheath knife all German soldiers carry. Before he could draw it clear and strike I seized his wrist with my left hand. He was a burly, powerful fellow, outweighing me at least fifty pounds. I was like a child in his grasp and it would be but a matter of seconds before he could free himself and plunge the knife into my ribs or neck.[43]

The German had attempted to keep Lynch close so that he could not use his advantage of distance, as Lynch was now the only one to still have a rifle with a fixed bayonet. Lynch had managed to ward off the imminent attack with the knife with his left hand, but, with his right hand still gripping his rifle, he was forced to use his legs, and one of the venerable traditions of *infighting*:

> But we had been trained for situations like this. Shifting my weight to my left foot I brought my right knee up with crushing force into his groin. At the paralyzing blow I felt his strength drain from him like water. With a mighty wrench I tore free of his failing grasp. A forward lunge and the point of my bayonet struck him in the breast and passed through to the hilt, leaving a few

inches of steel protruding from his back. He gave a hoarse scream of agony as the steel bit through his body, then collapsed.[44]

Kneeing or kicking an opponent in the groin had been advocated in official training literature since 1916 and in unofficial literature since 1915.[45] Having used his bayonet, Lynch was now left with the task of recovering his weapon:

> I attempted to withdraw the bayonet. It was caught fast in ribs and clothing. Panic seized me... I put one foot on the prostrate German and jerked frantically at the rifle, to no avail.
> My panic left as quickly as it came. Our training had also covered this emergency. Quickly throwing a cartridge into the chamber of my rifle, I pulled the trigger and sent the bullet crashing along the course of the bayonet. With a strong steady pull I withdrew the dripping blade. The bayonet was followed by a gush of blood that turned the field grey tunic of my late foe a dark red.[46]

Planting the foot and firing a round were both official solutions to the problem of a bayonet that had gotten stuck.[47] Lynch had not first resorted to the unofficial approach to the problem of twisting the bayonet. This suggests a level of some success for the Directorate of Bayonet Fighting and Physical Training in stamping out unauthorized practices.

Lynch's memoir offers a vivid individual perspective on a number of points relating to infantry tactics and close combat. The *dispersion* of German machine guns out of trench lines, the use of bounding by attacking infantry, and the 30-yard range of the bayonet charge: these were all significant elements of infantry fighting in the second half of the war. Lynch also recounted the practical application of several techniques of assault and bayonet training from 1917: firing from the hip, continuing the forward rush, and the knee to the groin. All these techniques were advocated by official training literature. Finally, Lynch's struggle to free his weapon demonstrated the success of the Directorate of Bayonet Fighting and Physical Training in combating unofficial techniques that had infiltrated the official training systems.

III. Arras: Continuing Patterns

Between 27 and 31 August, the Canadian Corps conducted daily operations that continued to bite into the Hindenburg Line.[48] The 1 and 2 September attacks by the 13th and 14th Battalions demonstrated the continued importance of conditioning and the negotiation of close combat. At 4:50 a.m. on the morning of 1 September, the 14th Battalion attacked behind a creeping barrage to secure jumping-off positions for an attack scheduled to take place the next morning. Their objective was Hans Trench, some 500 yards distant and used by the Germans as the center of an outpost zone in Drocourt-Queant Line. The battalion report on the attack commented on the quality of the German defenders: "The morale of the enemy was very poor. Never in my experience have I seen him so disorganized and thoroughly cowed. Given a barrage and infantry close on its heels, it was the easiest thing in the world to absolutely overrun him."[49] George Burdon McKean V.C., with the 14th Battalion, was involved in one of the bayonet charges that overran Hans Trench:

> Just then one of our officers came up. Although yelling at the top of his voice I barely heard him say: 'What about a bayonet charge?'
> 'Yes,' I shouted back, 'the very thing.'
> 'All right, I have a whistle here; when I blow it we'll all rush forward.'
> 'Righto, we'll be with you.'
> 'A bayonet charge,' I yelled, 'everybody for a bayonet charge.'
> 'A bayonet charge! a bayonet charge!' was the cry everywhere around me. Some of the boys laughingly felt the points of their bayonets.
> 'All ready,' I shouted.
> 'Yes, sir,' came back the reply.
> I signaled to the other officer – he blew his whistle and we made a wild rush forward, uttering the most weird, blood-curling cries. A solitary machine gun spat out for a few seconds, and two or three bombs were thrown from the trench, and then we were on top of them! The Huns made a brief effort at resistance, but it was short lived. Then they shrieked for mercy – but it was too late!"[50]

Although McKean's retelling of the event is certainly a little more plucky and sanitized than the reality of the assault, it provides an example of small groups of infantry charging after they had developed a sufficient mass from sections bounding forward in rushes behind the creeping barrage. Once close combat was offered, the Germans at Hans Trench decided, half-heartedly, to resist – however, as the Canadian troops closed the final distance resistance faltered and German troops tried to surrender. This decision to surrender came too late, and many were bayoneted. The battalion narrative of the operation commented specifically on the effect of close combat conditioning:

> A feeble attempt at resistance was attempted but my leading company had a large proportion of new drafts who, had joined the Battalion but a few hours before and were viewing their first Huns, went joyfully in with the bayonet and everyone ran wild for a few minutes, the Huns screaming for mercy – at least 50 were bayoneted.[51]

This assault resulted in an estimated 100 German killed.[52] Thus, according to the official report, the bayonet accounted for roughly 50% of German fatalities, in addition to some 300 Germans taken prisoner.[53] The bayonet charge had forced the decision on the defenders, and the half-hearted attempt to resist had resulted in a high ratio of bayonet kills.

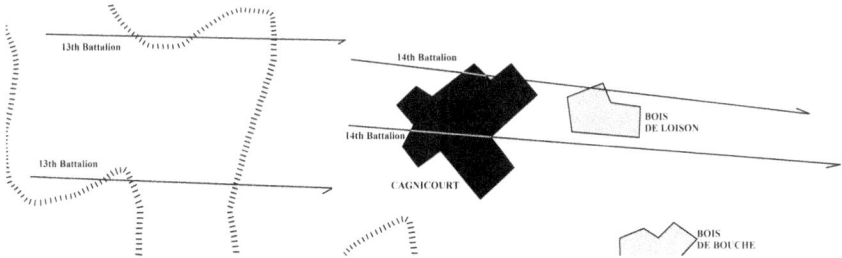

13th and 14th Battalion during the Battle of Arras, 2 September 1918

The next morning at 4:50 a.m., the 1st and 4th Canadian Divisions attacked hard on the heels of a creeping barrage. The advance of the 13th Battalion that morning demonstrated a number of elements of close combat. Jumping off from the positions captured by the 14th Battalion the previous day, the 13th Battalion pushed up the forward slope of a ridge roughly 1000 yards west of the village of Cagnicourt. Following the barrage closely, the battalion encountered little retaliatory fire and secured the first two trench lines on the forward slope. Before advancing over the crest of the ridge to the third German trench line, the leading companies were organized in waves, and the men of the 13th Battalion proceeded to clear the third trench "with bayonets fixed."[54] Major W. C. MacFarlane M.C., commanding the assaulting company, is recorded in the battalion diary as having bayoneted three Germans himself.[55]

Leap-frogging through the 13th Battalion at 8:00 a.m., the 14th Battalion advanced on the last line of fortifications in the Hindenburg Line, the Buissy Switch Trench, which lay roughly a mile and a half to the west. The Canadians cleared small copses and the village of Cagnicourt. In this terrain, the 14th Battalion encountered dispersed strong points which were "brushed away with ease"[56] by stalking these individual posts. The battalion diary described one of these small engagements in detail. "Shortly after jumping off, a part of No. 3 Company was held by an enemy machine gun, Lieut. A. L. McLean, M.C., D.C.M with a party of men was surrounding the gun with a view to rushing the position when one of the enemy walked in his direction with his hands up."[57] McLean and his men had enveloped and closed on the German position in order to rush it with the bayonet. The 14th Battalion narrative provided more detail on the way in which this encirclement was accomplished: "Lieut. McLean, M.C., D.C.M. commenced to encircle this nest and had succeeded in getting his Lewis gun to the right flank of it..."[58] Here, as was common practice in 1917 and 1918, Canadian infantry used their own firepower to suppress German strong points to cover the infantry as they closed for the final rush.

Once the Canadians had gathered sufficient numbers at close range and the bayonet charge threatened, the Germans began the negotiation of surrender. In this case it was only pretence: "Lieut. McLean rose to his feet and the Hun turned his gun on him and killed him instantly. His platoon, infuriated, rushed forward and killed every member of two crews and also killed a large number that were coming from CAGNICOURT to surrender."[59] Once the German machine gun crews had demonstrated the continued will to resist, conditioning took over, and the crew was rushed and dispatched. In their anger, the men of "C" Company continued to kill any German that appeared until there was a sufficient lull in the action. In spite of incidents such as this, the 13th and 14th Battalions succeeded in biting into the Buissy Switch Line and taking an estimated 1000 demoralized prisoners.[60]

The operations at Amiens, Scarpe, and Arras highlighted the many refinements in infantry and artillery tactics that had evolved over the course of 1917 and 1918; and yet, the bayonet continued to dominate the final phase of the infantry attack. The subjugation of new weapons like the tank to supporting infantry attacks demonstrated that new weapons were harnessed in order to assist rifle-and-bayonet-armed infantry in their ultimate task.

IV. Conclusion: Beyond the Canal du Nord

Throughout October, the instances of bayonet fighting diminished and this trend continued throughout the pursuit to Mons. As German morale faltered, so too did the willingness of Germans troops to resist, or even to remain long enough to surrender. After Arras, the Canadian Corps fought their way across the Canal du Nord through the Marcoing Line and into the city of Cambrai between 27 September and 11 October. A significant number of accounts of hand-to-hand fighting appear in the records of the fighting, but the number of accounts diminished in comparison to those recorded for Amiens, Scarpe, and Arras. The decrease in incidents of bayonet fighting was mirrored by a reduction in the number of prisoners taken by the advancing Canadians. When the Hundred Days is broken into the three periods of operations defined by the *Report of the Overseas Military Forces of Canada, 1918*, the daily average of the number of prisoners taken by the Canadian Corps dropped significantly: between 8 and 22 August 9,131 prisoners were taken or an average of 608.7 per day; between 23 August and 11 October, 18,585 prisoners were taken or 379.3 per day; and between 12 October to 11 November, 3,821 prisoners were taken or 123.3 per day.[61] After the fighting at the Canal du Nord, Canadians typically faced half-hearted attempts by German troops to delay the Canadian advance and the defenders often withdrew at the slightest threat. However, until the end, some Germans continued to resist the advance and Canadian troops were forced to attack these isolated pockets of resistance.[62] The bayonet continued to find a role up to the Armistice.

Conclusion

THE ANSWER TO the question posed by this work is straightforward: soldiers continued to fix bayonets before going "over the top" throughout the Great War because the bayonet was an important weapon in forcing the enemy to decide whether he would fight, flee, or surrender. The incidents catalogued and investigated in this book also indicate that this was not an infrequent occurrence; the cross-section of official records, battalion histories, and personal accounts indicate that the bayonet was used in nearly every battle and by every infantry formation of the Canadian Corps.

Bayonet fighting was an unpleasant and hazardous task – the tense wait after fixing bayonets for the order to go "over the top," crossing No Man's Land, closing on the enemy position, steeling the will to rush the final distance, and finally thrusting home with the bayonet. Soldiers did not enjoy it, or even the prospect of it. For critics 90 years removed, the bayonet charge still seems futile and pointless. Most historians have been content not to challenge the conclusion that the bayonet was obsolete in the Great War, and have, therefore, avoided serious investigation of this disquieting reality of the Western Front.

The comparison between the official records and the incidents of bayonet fighting in personal accounts and battalion histories demonstrates clearly that close combat was often hidden in the language of official records. In many cases, the use of the bayonet was implied by terms like "rushed,"[1] "swarmed,"[2] "successfully driven home,"[3] and "severe fighting."[4] In other cases, bayonet use was downplayed by the use of euphemistic terms, such as "successfully carried out,"[5] "cleared,"[6] "captured,"[7] and "secured,"[8] On occasion, some bayonet fighting was completely ignored by supposedly definitive language, such as "no fighting"[9] or "no opposition."[10] On other occasions, instances of bayonet fighting found in battalion histories and personal accounts were over-shadowed by descriptions of the predominant weapons or by tactical systems employed, "Lewis gun and rifle fire"[11] and "bombed"[12] are two such examples. However, while personal accounts and battalion histories provide a helpful adjunct to the official records, they too fail to document fully the close combat and bayonet fighting that occurred during the war. Many soldiers failed to leave records because they were unwilling, overwhelmed, or incapable after the experience of close combat.

Those records that exist often fail to highlight sufficiently the significant psychological impact of the bayonet. After the war, in an article entitled "Training in the use of the Bayonet," Lt. Col. G. Dalby emphasized the importance of the stress in bayonet fighting:

> We must seek for our answer by a study of human nature. We must look into the heart and soul of the man fighting in the forefront of the battle before we decide this question. We must find out what is at the back of the soldier's mind throughout the various stages of the attack, amidst the distractions of the battlefield... What passes through his mind when... he is nearing the enemy...?
>
> Surely at the back of every man's mind... must come the thought: *"What is going to happen when I come to close grips with my opponent?"* What will be the issue of such a situation? Shall I be able to knock him out or will he knock me out? Is not the fear of the assault or close combat the final and fundamental cause of troops leaving a position or surrendering when the enemy is actually occupying or on the point of taking that position?[13]

Dalby acknowledged the negotiation of close combat and the elementary question of the soldier's confidence in surviving the encounter. Actual close combat only occurred if the soldiers on both sides were roughly equal in confidence. This psychological impact was difficult to attribute to the bayonet in official records and thus frequently overlooked in historical investigation. Assaulting troops that lacked sufficient weight of numbers after crossing No Man's Land might fail to launch or complete the bayonet charge, and be forced back. If the assaulting troops had sufficient weight in numbers, the bayonet charge, or at least the threat of it, forced the defenders to decide whether to fight, flee, or surrender.

However, there are records of this psychological dimension, and they provide additional – and largely unappreciated – evidence of the bayonet's importance on the Western Front. The recriminations and blame for an attack that failed, or comments that soldiers had to withdraw after exhausting the supply of bombs, document the stresses assaulting troops had to overcome and on which shock tactics relied. Conversely, the surrender or flight of the enemy adds a significant, but unquantifiable, effect of the bayonet – even if only in an assisting role. The bayonet was used frequently on the Western Front, and, given the synthesis of official, personal, and historical accounts of the war, it was a more frequent occurrence than quantifiable numbers of isolated accounts, casualty lists, or wounding surveys are capable of indicating.

Dalby also observed the threat of close combat as the "fundamental" – or more accurately, immediate – "cause" of the enemy fleeing from or capitulating to attacking troops. Dalby, writing in defence of the bayonet, perhaps overstated the case and failed to acknowledge that all the weapons and tactics in the offensive arsenal played an important and cooperative role in the attack. Yet, the importance of the threat of close combat in clearing enemy positions was continually reiterated by the continuity of the bayonet charge in Canadian attacks throughout the war. What had changed was the nature of the support provided by other arms and weapons in the infantry attack. In 1914, the infantry had rifles, machine guns, and direct fire of field guns to suppress enemy weapons. In 1918, grenades (hand and rifle), light machine guns, mortars, howitzers, smoke, gas, creeping barrages, counter-battery fire, aircraft, and tanks had all been introduced and harnessed to assist the infantry forward. There were considerable changes in weapons and tactics during the war, but the basic principles of the attack remained unchallenged. Thus, in 1918, just as in 1914, fire neutralized enemy fire, permitting attacking troops to close on enemy positions in order to assault with the bayonet.

Hindsight has made it clear that armies inclining too much toward the *élan* faced horrendous casualties in the face of modern firepower in the first years of the war. A more difficult point to appreciate was that crawling forward, throwing a handful of bombs, and then withdrawing would not eject the enemy from advantageous positions either. If ground was to be taken, soldiers had to be sufficiently trained and conditioned to be confident in the use of the bayonet. Bayonet training permitted soldiers to overcome the stress of battle, intimidate defending troops, and engage in hand-to-hand fighting if the enemy did not break. Dalby concluded:

> If it be agreed that close combat or the threat of the assault on the part of the infantry… is necessary before decisive victory can be won, surely it would be most unwise not to fortify the heart of the soldier for the close combat fight; and the bayonet, in a greater or lesser degree depending on circumstances, is

the necessary adjunct to the bullet for close combat. Therefore we say, 'Let us give the soldier confidence in the use of the bayonet.'

He may never use it, he may not be required to use it, but, on the other hand, he may have to use it, and in a hard fought fight will always think that he is going to use it.[14]

Preparing soldiers for the "circumstance" of close combat alone justified training in bayonet fighting and close combat. Additional justification is found in the psychological impact of the bayonet, which cannot be overstated. The bayonet charge itself had a tremendous effect, even if a soldier did not physically engage the enemy in hand-to-hand fighting. William Patrick Doolan, a veteran with the 21st Battalion, and later a professor at Queen's University, summarized this imperative to close with the enemy in an interview for CBC Radio's 1964 production of *In Flanders Fields*: "the main function of an infantryman is to... come in contact with the enemy and drive him back. That is his objective, you can't win a war without coming in contact with people you are fighting."[15] Making contact often drove the enemy to surrender or break – however, merely making contact was often not enough. If the enemy was not intimidated, soldiers were forced to fight hand-to-hand. And for making contact with the enemy, the bayonet proved a capable weapon in the trenches of the Western Front.

Abbreviations

AGS	Army Gymnastic Staff (British)
BEF	British Expeditionary Force
Bde	Brigade
Btn	Battalion
CAGS	Canadian Army Gymnastic Staff
CEF	Canadian Expeditionary Force
CMR	Battalion, Canadian Mounted Rifles
HE	High Explosive
HMSO	His/Her Majesty's Stationary Office
NCO	Non-Commissioned Officer
PPCLI	Princess Patricia's Canadian Light Infantry
RCR	Royal Canadian Regiment
RG	Record Group (Library and Archives Canada)
WD	War Diary (Library and Archives Canada)

Endnotes

Introduction

[1] Timothy Bowman, *The Irish Regiments in the Great War: Discipline and Morale* (New York: Manchester University Press, 2006), 159.

[2] John Howard Morrow and J. Morrow Jr., *The Great War: An Imperial History* (London: Routledge, 2005), 42; Timothy Woods, Andrew A. Wiest and M. K. Barbier, *Strategy and Tactics, Infantry Warfare* (St. Paul: Zenith Imprint, 2002), 16.

[3] Ian F. W. Beckett, *The Great War 1914-1918* (New York: Pearson Education, 2001), 42; Pierre Berton, *Vimy* (Toronto: Anchor, 2001), 117; Shelford Bidwell and Dominick Graham, *Firepower, British Weapons and Theories of 1904-1945* (Boston: George Allen & Unwin, 1982), xv; Brian Bond, *War and Society in Europe, 1870-1970* (Montreal: McGill-Queen's University Press, 1998), 92 and 101; Max Boot, *War Made New: Technology, Warfare, and the Course of History, 1500 to Today* (New York: Gotham, 2006), 151; D. J. Goodspeed, *The Road Past Vimy: The Canadian Corps 1914-1918* (Toronto: MacMillan of Canada, 1969), 53; Archer Jones, *The Art of War in the Western World* (University of Illinois Press, 2001), 626; Mary R. Habeck, "Technology in the First World War: The View from Below," *The Great War and the Twentieth Century*, Ed. Jay Winter, Geoffrey Parker, and Mary R. Habeck (New Haven: Yale University Press, 2000), 100; Bill Rawling. *Surviving Trench Warfare: Technology and the Canadian Corps 1914-1918.* (Toronto: University of Toronto Press, 1992), 11.Martin Samuels, *Command or Control? Command, Training and Tactics in the British and German Armies, 1888-1918* (Portland: Frank Cass, 1995), 116; Jay Stone and Erwin A. Schmidl, *The Boer War and Military Reforms* (Lanham: University Press of America, 1988), 14; A. J. P. Taylor, *Illustrated History of the First World War* (New York: G. P. Putnam's Sons, 1964), 44-5; Tim Travers, "Learning and Decision-Making on the Western Front, 1915-1916: The British Example," *Canadian Journal of History* 18, No. 1 (April 1983), 96-7; ---, *The Killing Ground The British Army, the Western Front and the Emergence of Modern Warfare 1900-1918* (London: Allen & Unwin, 1987), 86-9; Jay Winter and Blaine Baggett, *The Great War And the Shaping of the 20th Century* (New York: Penguin Studio, 1996), 59-70; This tendency has also observed by John Lee, "Some Lessons of the Somme: The British Infantry in 1917," *Look to your front, Studies in the First World War*, Brian Bond et al. (Staplehurst: Spellmount, 1999), 79; Gary Sheffield, *Forgotten Victory, The First World War: Myths and Realities* (London: Headline, 2001) xiii and 87.

[4] Tim Travers, "The Offensive and the Problem of Innovation in British Military Thought 1870-1915," *Journal of Contemporary History* 13 (1978), 531.

[5] Paul Hodges, "'They don't like it up 'em': Bayonet fetishization in the British Army during the First World War." *Journal of War and Culture Studies* 1, No. 2, 127.

⁶ Jacob Lee Hamric, "Germany's Decisive Victory: Falkenhayn's Campaign in Romania, 1916" (M.A. Thesis: Eastern Michigan University, 2004), 11; Morrow and Morrow, *The Great War*, 31; Nicholas A. A. Murray, "The Theory and Practice of Field Fortification from 1977-1914" (P.h.D. Dissertation, St. Anthony's College, University of Oxford, 2007), 218; Tim Travers, *The Killing Ground, The British Army, the Western Front and the Emergence of Modern Warfare 1900-1918.* (London: Allen & Unwin, 1987), 37-61; J. M. Winter, *The Experience of World War I*, (Oxford University Press, 1989), 85; Winter and Baggett, *The Great War,* 59-60; Joseph C. Arnold, "French Tactical Doctrine, 1870-1914," *Military Affairs* 42, No. 2 (April 1978), 88; Brian Bond, *War and Society in Europe,* 92. Tim Cook, *At the Sharp End, Canadians Fighting the Great War 1914-1916, Volume One* (Toronto: Viking Canada, 2007), 12; Richard F. Hamilton and Holger H. Herwig, *The Origins of World War I* (New York: Cambridge University Press, 2003), 253.

⁷ Joanna Bourke, *An Intimate History of Killing, Face-to-Face Killing in Twentieth century Warfare* (London: Granta, 1999), 54 and 92-3; William C. Fuller Jr., "What is a Military Lesson?," *Strategic Logic and Political Rationality: Essays in Honor of Michael I. Handel*, Ed. Bradford A. Lee and Karl-Friedrich Walling (London: Routledge, 2003), 43; I. B. Holley Jr., *Technology and Military Doctrine, Essays on a Challenging Relationship* (Maxwell: Air University Press, 2004), 69.

⁸ Mark Connelly, *Steady the Buffs!: Regiment, a Region, and the Great War* (New York: Oxford University Press, 2006), 53; John Laffin, *British Butchers and Bunglers of World War One* (Gloucester: Alan Sutton, 1988), 11; Lyn Macdonald, *1915: The Death of Innocence* (Baltimore: John Hopkins University Press, 2000), 14, 96, 102, 204, 208, and 373; Desmond Morton, "Changing operational Doctrine in the Canadian Corps 1916-1917," *The Army Doctrine and Training Bulletin* 2, No. 4 (Winter 1999), 35.

⁹ "ACI 1103: Training of Infantry Recruits," 31 May 1916 and "ACI 1968: Training of Category A Infantry Recruits," 15 October 1916, War Office, *Army Council Instructions 1916* (London: Harrison and Sons, 1916). War Office; *Infantry Training 1914* (London: Harrison & Sons, 1915), Appendix 2.

¹⁰ Stephen J. Harris, *Canadian Brass: The Making of a Professional Army, 1860-1939* (Toronto: University of Toronto, 1988), 75.

¹¹ Mark Osborne Humphries, "'Old Wine in New Bottles,' A Comparison of British and Canadian Preparations for the Battle of Arras," *Vimy Ridge, A Canadian Reassessment*, Ed. Geoffery Hayes, Andrew Iarocci, and Mike Bechthold (Waterloo: Wilfrid Laurier University Press, 2007), 80; Gary Sheffield, "Vimy Ridge and the Battle of Arras: A British Perspective," *Vimy Ridge, A Canadian Reassessment,* 23.

¹² W. G. Clifford, *The British Army* (Alcester: Read Books, 2008), 58-59.

[13] Tim Cook, *At the Sharp End*, 319; Antulio Joseph Echevarria, *Imagining Future War: The West's Technological Revolution and Visions of Wars to Come, 1880-1914* (Westport: Greenwood Publishing Group, 2007), 33.

[14] Various Authorities, *Canada in the Great World War: An authentic account of the Military History of Canada from the earliest days to the close of the war of the Nations. Volume VI: Special Services, Heroic Deeds, Etc.* (Toronto: United Publishers of Canada), 270-312.

[15] *Canada in the Great World War*, 270-312.

[16] LAC RG 41, vol. 9, 16th Btn., Mr. B. C. Lunn, 2/8.

[17] Ibid, 2/8.

Chapter I: Myths and Misconceptions

[1] Paddy Griffith, *Battle Tactics*, 3.

[2] Dan Todman, *The Great War, Myth and Memory* (New York: Hambledon, 2005), 20.

[3] Brian Bond, *The Unquiet Western Front, Britain's role in literature and history* (New York: Cambridge University Press, 2002), 42-9; John Keegan, *The First World War* (New York: Viking, 1998), 311; Albert Palazzo, *Seeking Victory on the Western Front: The British Army and Chemical Warfare in World War 1* (Lincoln: University of Nebraska Press, 2000), 3; Hunt Tooley, *The Western Front, Battle Ground and Home Front in the First World War* (New York: Palgrave, 2003), 156.

[4] A. J. P. Taylor's description of the Second World War as "a good war" in comparison to the First World War is cited in: Bond, *The Unquiet Western Front*, 63; Sheffield, *Forgotten Victory*, xiii; Todman, *The Great War,* 8, 135.

[5] Gary J. Cox, "Of Aphorisms, Lessons, and Paradigms: Comparing the British and German Official Histories of the Russo-Japanese War," *The Journal of Military History* 56 (April 1992), 391; Alan Kramer, *The Dynamic of Destruction: Culture and Mass Killing in the First World War* (New York: Oxford University Press, 2007), 212.

[6] John Harris, *The Somme: Death of a Generation* (London: White Lion, 1966), 107.

[7] Beckett, *The Great War,* 462-5.

⁸ See: Bond, *The Unquiet Western Front* and Todman, *The Great War, Myth and Memory*.

⁹ Original emphasis, John Ellis, *Eye-Deep in Hell* (London: Croom Helm, 1976), 82.

¹⁰ Harris, *The Somme*, 35

¹¹ For "the learning curve" in the Canadian historiography of the Great War, see: Marc Osborne Humphries, "The Myth of the Learning Curve, Tactics and Training in the 12th Canadian Infantry Brigade, 1916-1918," *Canadian Military History* 14, No. 4 (autumn 2005), 15.

¹² Tim Cook, *Clio's Warriors, Canadian Historians and the Writing of the World Wars* (Vancouver: UBC Press, 2006), 208-9; John Grodzinski, "The Use and Abuse of the Battle: Vimy Ridge and the Great War over the History of the First World War," *Canadian Military Journal* Vol. 10, No. 1, 83-6; Andrew Iarocci, *Shoestring Soldiers: The 1st Canadian Division at War, 1914-1915* (Toronto: University of Toronto Press, 2008), 275.

¹³ Vance, *Death So Noble*, 172.

¹⁴ Desmond Morton and J. L. Granatstein, *Marching to Armageddon: Canadians and the Great War 1914-1919* (Toronto: Lester & Orpen Dennys, 1988), caption for H. J. Mowat, 'Trench Fight," between 32 and 33; Desmond Morton, *When Your Number's Up: The Canadian Soldier in the First World War* (Toronto: Random House of Canada, 1993), 284, note 58; Kevin R. Shackleton, *Second to None: The Fighting 58th Battalion of the Canadian Expeditionary Force* (Toronto: Dundurn Group, 2002), 266; Rawling, *Surviving Trench Warfare*, 68; William Frederick Stewart, "Attack Doctrine in the Canadian Corps, 1916-1918" (M.A. Thesis, University of New Brunswick, 1982), 60.

¹⁵Tim Ripley, *Bayonet Battle: Bayonet Warfare in the 20th Century* (London: Sidgwick & Jackson, 2002), 255.

¹⁶ Ibid, 5.

¹⁷ Ibid, 44.

¹⁸ Hodges, "'They don't like it up 'em'," 124; also found in Bourke, *An Intimate History of Killing*, 54.

¹⁹ Hodges, "'They don't like it up 'em'," 123.

²⁰ Also found in Bourke, *An Intimate History of Killing*, 231.

²¹ Hodges, "'They don't like it up 'em'," 134.

²² David Grossman, *On Killing: The Psychological Cost of Learning to Kill in War and Society* (Boston: Back Bay Books, 1996), 122; Paul Hodges, "'They don't like it up 'em'," 127; John Keegan, *The Face of Battle: A Study of Agincourt, Waterloo and the Somme* (New York: A. Knopf, 1976), 264; Kramer, *The Dynamic of Destruction,* 252; T. H. McGuffie, "The Bayonet: A survey of the weapon's employment in warfare over the past three centuries," *History Today,* vol. 12 (August 1962), 593; Leonard V. Smith, Stéphane Audoin-Rouzeau, Annette Becker, *France and the Great War, 1914-1918* (New York: Cambridge University Press, 2003), 93; Denis Winter. *Death's Men: Soldiers of the Great War* (Markham: Penguin Books, 1978), 40.

²³ Hodges, "'They don't like it up 'em'," 125; McGuffie, "The Bayonet," 593; O'Leary, "À la bayonet or Hot Blood and Cold Steel," *Canadian Army Infantry Journal* (Spring 2000), 8; Denis Winter, *Death's Men,* 110.

²⁴ J. M. Winter, *The Experience of World War I,* 122. Rawling, *Surviving Trench Warfare,* 68. Morton, *When Your Numbers up,* 284. Frederick P. Todd, "The Knife and Club in Trench Warfare, 1914-1918," *The Journal of the American Military History Foundation,* 2, No. 3 (Autumn 1938), 140. Grossman, *On Killing,* 123.

²⁵ McGuffie, "The Bayonet," 593; Denis Winter, *Death's Men,* 40 and 110; M. M. O'Leary, "À la bayonet," 8; John Ellis, *The Social History of the Machine Gun* (Baltimore: Johns Hopkins University Press, 1986), 126; Rawling, *Surviving Trench Warfare,* 68; Morton, *When Your Number's Up,* 284.

²⁶ T. J. Mitchell and G. M. Smith, *Medical Services: Casualties and Medical Statistics of the Great War* (HMSO, 1931), 40; Cited in: Denis Winter, *Death's Men,* 110; Tim Cook, *Shock Troops, Canadians Fighting the Great War 1917-1918, Volume 2* (Toronto: Viking Canada, 2008), 614.

²⁷ Harry L. Gilchrist, *A Comparative Study of World War Casualties: from Gas and Other Weapons* (Washington: U. S. Government Printing Office, 1928), 19; Cited in A. D. Harvey, "The Bayonet in Battle," *RUSI Journal* 150, No. 2 (April 2005), 62.

²⁸ McGuffie, "The Bayonet," 593.

²⁹ Denis Winter, *Death's Men,* 110; and Nicholas Murray, "The Theory and Practice of Field Fortification," 160 note 142.

³⁰ Department of Militia and Defence, *The Organization of Bayonet Fighting and Physical Training in a Battalion C.E.F. (Revised) 1916* (Ottawa: Government Printing Bureau, 1916), 18; Department of Militia and Defence, *Bayonet Training 1916 (Provisional)* (Government Printing Bureau, 1916), para. 23.

[31] *Bayonet Training 1916*, 1

[32] LAC RG 9 vol. 3842, 43/13-14, "Summary of Operations, 8th Brigade, 11 to 17 September," LAC RG 9 III-C-3, vol. 4016, Folder 31, File 7, "Operations Scarpe and Drocourt-Queant Line (3rd Cdn. Inf. Brigade) 26-8-18 to 2-9-18," 3.

[33] Cook, "The Politics of Surrender," 641.

[34] LAC RG 9, III-D-3, vol. 4944, LAC RG 9, III-D-3, vol. 4944, WD, 85th Btn., June 1917, Appendix B, "Summary of Operations June 25th to July 1st 1917;" LAC RG 9 III-C-3, vol. 4229, Folder 19, File 5, "Operations Avion 12th CIB," S.G. 23/866, "Report on Operations carried out by the 12th Canadian Infantry Brigade between 26th June and 28th June 1917."

[35] Denis Winter, *Death's Men*, 110.

[36] McGuffie, "The Bayonet," 593.

[37] R. C. Featherstonhaugh, *The Royal Montreal Regiment 14th Battalion, C. E. F. 1914-1925*, (Montreal: Gazette Publishing, 1927), 53.

[38] Harold R. Peat, *Private Peat* (Indianapolis: Bobbs-Merrill, 1917), 94; LAC RG 41, vol. 13, 42nd Btn., J. M. Morris, 1/14.

[39] Rat hunting incidents are found in: Harold Baldwin, *Holding the Line* (Toronto: George J. McLeod, 1918) 192-3; Shackleton, *Second to None*, 33; Bert Drader, 26 June 1916, From Canadian Letters & Images Project; LAC RG 41, vol. 15, 50th Btn., H. J. Pattison, 1/10-11; LAC RG 41, vol. 16, 72nd Btn., E. Plumsteel, 1/9; As well, one incident of bayonets used on moles at a Battalion farm is found in LAC RG 41, vol. 17, 4th CMR, Gregory Clark, 1/11.

[40] William D. Matheson. *My Grandfather's War: Canadians Remember the First World War 1914-1918* (Toronto: Macmillan of Canada, 1981), 43.

[41] Baldwin, *Holding the Line*, 192-3.

[42] Cited in Morton, *When Your Number's Up*, 119.

[43] D. G. Scott Calder, *The History of the 28th (Northwest) Battalion, C.E.F. (October 1914-June 1919)*, (Unpublished History, Fort Frontenac Library), 47; Hodges, "'They don't like it up 'em'," 125; "Todd, "Knife and Club," 140.

[44] Anon, *Letters From the Front: Being a Record of the art played by Officers of the Bank in the Great War 1914-1919; Volume 1*. Ed. Charles Lyons Foster (Toronto: Southam Press, 1920), 86-7; Kim Beattie, *48th Highlanders of Canada, 1891-1928*

(Toronto: Southam Press, 1932), 206; Joseph Chabelle, "Courcelette: The Glorious Battle Fought by the 22nd French Canadian Regiment," La Canadienne, October 1920, [original in French, translated copy at CWM], 14a. cited in Cook, *At the sharp end*, 460-1; R. C. Featherstonhaugh, *The 24th Battalion, C.E.F., Victoria Rifles of Canada 1914-1919* (Montreal: Gazette Printing, 1930),191; —, *The 13th Battalion Royal Highlanders of Canada: 1914-1919* (Canada: The 13th Battalion, Royal Highlanders of Canada, 1925), 103-4; Joseph Hayes, *The Eighty-Fifth in France and Flanders* (Halifax: Royal Print & Litho, 1920), 59; Iarocci, *Shoestring Soldiers*, 149; Gordon Reid. *Poor Bloody Murder: Personal Memoirs of the First World War*, (Oakville: Mosaic Press, 1980), 67; Shackleton, *Second to None*, 121; LAC RG 9 III-C-3, vol. 4017, Folder 34, file 9 Operations Minor (10th Bn) "Report of Minor Offensive Operation carried out on night of 4-5th February 1916 by 10 Canadian Infantry Battalion;" LAC RG 41, vol. 8, 5th Btn., R. L. Christopherson, 2/11; LAC RG 41, vol. 8, 10th Btn. William Walkinshaw and C. Scriven, 3/10; LAC RG 41, vol. 10, 21st Btn., A. R. Cousins, 1/ 17; LAC RG41, vol. 11, 26th Btn., Mr. Ingram 5/3; LAC RG 41, vol. 14, 46th Btn., Crowe, 2/10; LAC RG 41, vol. 16, 102nd Btn., Roy Gross, 1/12; LAC RG 41, vol, 17. 2nd CMR, Gus Siverts, 2/3.

[45] Author's emphasis. LAC RG 9, III-D-3, vol. 4867, WD, 1st Bde., December 1916, Appendix C, "OPERATION ORDER NO. 36," 20.

[46] Bill Rawling, *Surviving Trench Warfare*, 172.

[47] Ibid, 17-20.

[48] Alexander McClintock, *Best O' Luck: How a Fighting Kentuckian won the Thanks of Britain's King* (Vernon: CEF Books, 2000), 32; LAC RG 9 III 3108 T-5-36. LAC, RG 41, vol. 7, 1st Battalion, M. C. McGowan, 1/1. "And we used to have to make our jam pot bombs which were two ounces of cotton and one ounce of dry cotton primer and oinide of mercury detonator, with a four second time fuse. We used to have to make them, carry them in jackets… nevertheless you just started out with bayonet men and so forth that led you and you were suppose to clear the trench by throwing these bombs." LAC RG 41, vol. 15, 50th Battalion, Mr A. Turner, 4/2:
   "And it was my job to throw mills bombs ahead of the bayonet men…"

[49] RG 41, volume 14 47th Battalion Mr. Ormond St. Patrick Aitkens Tape 1, page 9

[50] An example of this being used in training is found in: A. Fortescue Duguid, *The Official History of Canadian Forces in the Great War: Volume 1* (Peternaude, 1938), *Part I*, 175 and *Part II*, 199-200.

[51] *The Organization of Bayonet Fighting*, 22.

[52] Headquarters Canadian Overseas Military Forces, *Bayonet Fighting Illustrated 1917* (London: Harrison and Sons, 1917), 86-7. Training in knife fighting is also advocated by *The Organization of Bayonet Fighting...*,22.

[53] Grossman, *On Killing*, 122.

[54] Ibid, 123.

[55] Ibid, 123.

[56] Ibid, 123.

[57] Angelo Viggiani, *Lo Schermo*, (Venice: Giorgio Angeleri, 1575), verso 52.

[58] *Bayonet Training 1916*, para. 23.

[59] Grossman, *On Killing*, 124.

[60] This importance of training in overcoming the adrenaline of close combat is observed by Bourke, *An Intimate History of Killing*, 72.

[61] *Infantry Training 1914*, appendix 1, section 5. Similar admonishments are found in *Infantry Training 1911*, appendix 1, section 2; *Bayonet Training 1916*, para. 32.

[62] Charles W. Sanders Jr., *No Other Law: The French Army and the Doctrine of the Offensive* (Santa Monica: Rand Corporation, 1987), 27.

[63] Michael Howard, "Men Against Fire: Expectation of War in 1914," *International Security* 9, No 1 (Spring 1984), 57.

[64] J. M. Winter, *The Experience of World War I*, 126; Dennis E. Showalter, "Prussia, Technology and War: Artillery from 1815 to 1914," *Men, machines & war*, Ed. Ronald Haycock and Keith Neilson (Waterloo: Wilfrid Laurier University Press, 1988), 142; Edward G. Lengel, *To Conquer Hell: The Meuse-Argonne, 1918 The Epic Battle That Ended the First World War* (New York: Lippincott Williams & Wilkins, 2009), 22; Holley, *Technology and Military Doctrine*, 31; Henry G. Gole and William A. Stofft, *General William E. DePuy: Preparing the Army for Modern War* (Lexington: University Press of Kentucky, 2008), 14.

[65] Smith, Audoin-Rouzeau, and Becker, *France and the Great War*, 20; Robert B. Bruce, *Petain: Verdun to Vichy* (Dulles: Brassey's, 2008), 19.

[66] Tim Cook, *No Place to Run, The Canadian Corps and Gas Warfare in the First World War* (Toronto: UBC Press, 1999), 16; Hamric, "Germany's Decisive Victory," 11; Morrow and Morrow, *The Great War*, 31; Nicholas Murray, "The

Theory and Practice of Field Fortification," 218; Travers, *The Killing Ground*, 37-61; J. M. Winter, *The Experience of World War I*, 85; Winter and Baggett, *The Great War,* 59-60; Arnold, "French Tactical Doctrine," 88; Brian Bond, *War and Society in Europe,* 92.

[67] Cook, *At the Sharp End,* 12; Hamilton and Herwig, *The Origins of World War I* 253.

[68] David F. Burg and L. Edward Purcell, *Almanac of World War I* (Lexington: University Press of Kentucky, 2004), x; Morton, *When Your Number's Up*, 149; Robert A. Doughty, *Pyrrhic Victory: French Strategy and Operations in the Great War* (Cambridge: Harvard University Press, 2005), 28 and 75; Laffin, *British Butchers and Bunglers,* 10-11; Lengel, *To Conquer Hell,* 22; Morrow and Morrow, *The Great War,* 68; Louis Decimus Rubin, *The Summer the Archduke Died: On Wars and Warriors* (University of Missouri Press, 2008), 23; John Terraine, *The Great War* (London: Wordsworth Editions, 1997), 25 and 140: Terence Zuber, *The Battle of the Frontiers: Ardennes 1914* (Chalford: Tempus, 2007), 78.

[69] David G. Herrmann, *The Arming of Europe and the Making of the First World War* (Princeton: Princeton University Press, 1996), 223.

[70] Rawling, *Surviving Trench Warfare,* 68.

[71] Morton, *When Your Number's Up*, 284.

[72] *Infantry Training 1914*, Chapter X, Section 121, para 6-7. The instruction of small groups of infantry to bound or use rushes to approach the enemy are also found in: War Office, *Infantry Training 1902* (London: Eyre & Spottiswood, 1902), Part IV, section 220-1; ——, *Infantry Training 1905* (London: Wyman and Sons, 1908), Part IV, section 136; ——, *Infantry Training 1911* (London: Wyman and Sons, 1911), Part IV, section 125. It is also observed in: Cook, *At the Sharp End*, 85; M. A. Ramsay, *Command and Cohesion, The Citizen Soldier and Minor Tactics in the British Army, 1870-1918* (Westport: Praeger, 2002), 161.

[73] *Infantry Training 1914*, chapter X, Section 121, paras. 6-7.

[74] PRO WO158 344; cited in Griffith. *Battle Tactics*, 67; Ramsay, *Command and Cohesion,* 179; Samuels, *Command or Control?,* 109-110.

[75] LAC RG 9 III-C-3, vol. 4031, Folder 26, file 7, "Training: 1st Canadian Infantry Brigade," G.B./165, 10 May 1917.

[76] *Infantry Training 1911*, appendix 1, section 9.

[77] *Infantry Training 1914*, appendix 1, section 9. A pace is defined in *Infantry Training 1914* as "In *slow* and *quick time* the length of a pace is 30 inches. In *stepping out*, it is 33 inches, in *double time*, 40..." *Infantry Training 1914*, Chapter 2, section 19, subsection 1.

[78] Original emphasis. LAC RG 9 III, 3870, Folder 112, File 11,

[79] Ramsay, *Command and Cohesion,* 175.

Chapter II: 1870 to 1914, The Bayonet Before the War

[1] Carl von Clausewitz, *On War*, tr. Peter Paret (Princeton: Princeton University Press, 1976), 83-4.

[2] Antulio J. Echevarria II, "The 'Cult of the Offensive' Revisited: Confronting Technological Change Before the Great War," *Journal of Strategic Studies*, Vol. 23, No. 1, (March 2002), 201.

[3] Ramsay, *Command and Cohesion,* 79; Iarocci, *Shoestring Soldiers*, 14; Brian Bond, *War and Society in Europe,* 17-18.

[4] Ramsay, *Command and Cohesion,* 79. Iarocci, *Shoestring Soldiers,* 14.

[5] Bond, *War and Society in Europe,* 17.

[6] Beckett, *The Great War*, 43.

[7] Stone and Schmidl, *The Boer War,* 19, 80, and 82.

[8] Howard, "Men Against Fire," 49-50; Echevarria, *After Clausewitz*, 42, 104.

[9] Ramsay, *Command and Cohesion,* 159; Stone and Schmidl, *The Boer War,* 108.

[10] Joel A. Setzen, "Background to the French Failures of August 1914: Civilian and Military Dimensions," *Military Affairs* 42, No. 2 (April 1978), 87.

[11] Herrmann, *The Arming of Europe,* 188-9.

[12] Stone and Schmidl, *The Boer War,* 10, 121, 134, 140, and 149; Bidwell and Graham, *Firepower,* 39; Alex D. Haynes, "The Development of Infantry Doctrine in the Canadian Expeditionary Force," *Canadian Military Journal* (Autumn 2007), 66; Herrmann, *The Arming of Europe,* 59; Travers, "The Offensive and the Problem of Innovation," 533 and 545.

[13] Boot, *War Made New*, 9.

[14] Brian Bond, *War and Society in Europe*, 32; Herrmann, *The Arming of Europe*, 11; Roman Johann Jarymowycz and Donn A. Starry, *Cavalry from Hoof to Track* (Westport: Greenwood Publishing Group, 2008), 125; Rob Engen, "Steel against Fire: The bayonet in the First World War," *Journal of Military and Strategic Studies* 8, No. 3 (Spring 2006), 6,

[15] Hew Strachan, *The First World War, Volume 1: To Arms* (New York: Oxford University Press, 2001), 188.

[16] Antulio J. Echevarria II, *After Clausewitz, German Military Thinkers Before the Great War*, (Lawrence: University Press of Kansas, 2000), 94.

[17] Grandmaisson quoted in Hoffman Nickerson, *The Armed Horde*, (New York, 1940), 224.

[18] Grandmaisson quoted in Corelli Barnett, *The Sword Bearers, Studies in Supreme Command in the First World War*, (Toronto: Hodder & Stoughton, 1986), 247.

[19] Tim Travers, *The Killing Ground*, 37.

[20] Morrow and Morrow, *The Great War*, 31

[21] Beckett, *The Great War*, 43.

[22] Stone and Schmidl, *The Boer War*, 118. Samuels, *Command or Control?*, 62 and 75.

[23] Barnett, *The Sword Bearers*, 245-8; Arnold, "French Tactical Doctrine," 62, 64, and 65; Echevarria, *After Clausewitz*, 59, 70-73, 96

[24] Echevarria, *After Clausewitz*, 24, 33-5, 104. Travers, "The Offensive and the Problem of Innovation," 531.

[25] Griffith, *Battle Tactics*, 49-50.

[26] Michael Howard, "Europe on the Eve of the First World War," *The Coming of the First World War*, Ed. Robert John Weston Evans, Hartmut Pogge von Strandmann (New York: Oxford University Press, 1991); Connelly, *Steady the Buffs!*, 41; Arnold, "French Tactical Doctrine," 61.

[27] Arnold, "French Tactical Doctrine," 61; Howard, "Men Against Fire" 50.

[28] Arnold, "French Tactical Doctrine," 62.

[29] Howard, "Men Against Fire," 50.

[30] Ibid, 50.

[31] Samuels, *Command or Control?*, 71.

[32] David Stevenson, *Cataclysm, The First World War as Political Tragedy* (New York: Basic Books, 2004), 18-19.

[33] Kenneth Radley, *We Lead Others Follow: First Canadian Division, 1914-1918* (St. Catharines: Vanwell, 2006) 47.

[34] Stone and Schmidl, *The Boer War*, 40.

[35] Ibid, 82.

[36] Howard, "Men against Fire," 46; Stone and Schmidl, *The Boer War*, 36.

[37] Rawling, *Surviving Trench Warfare*, 9-10.

[38] Stone and Schmidl, *The Boer War*, 79 and 115; Travers, "The Offensive and the Problem of Innovation," 537.

[39] Stone and Schmidl, *The Boer War*, 95.

[40] Herrmann, *The Arming of Europe*, 96; Stone and Schmidl, *The Boer War*, 95-6.

[41] Dominick Graham, "Observations of the Dialectics of British Tactics, 1904-45," from *Men Machines & War*, Ed. Ronald Haycock and Keith Neilson (Waterloo: Wilfrid University Press, 1988), 54; George H. Cassar, *The Tragedy of Sir John French* (Mississauga: University of Delaware Press, 1985), 62; Stone and Schmidl, *The Boer War*, 117.

[42] Travers, "The Offensive and the Problem of Innovation," 537.

[43] Rawling, *Surviving Trench Warfare*, 10; Stone and Schmidl, *The Boer War*, 81 and 116.

[44] Bruce I. Gudmundsson, *Stormtroop Tactics, Innovation in the German Army, 1914-1918* (New York: Praeger, 1989), 20-21; Herrmann, *The Arming of Europe*, 87.

[45] Arnold, "French Tactical Doctrine," 63; Herrmann, *The Arming of Europe*, 81.

[46] Bond, *War and Society in Europe*, 84-85.

[47] Ripley, *Bayonet Battle,* 19-21.

[48] Ian Frederick William Beckett and Keith Simpson, *A Nation in Arms: A Social Study of the British Army in the First World War* (Dover: Manchester University Press, 1985), 47; Cook, *At the Sharp End*, 11; Rawling, *Surviving Trench Warfare,* 11; Cox, "Of Aphorisms, Lessons, and Paradigms," 393 and 397; Ramsay, *Command and Cohesion,* 150; Herrmann, *The Arming of Europe,* 22; Travers, "The Offensive and the Problem of Innovation," 537.

[49] Travers, "The Offensive and the Problem of Innovation," 538.

[50] Engen, "Steel against Fire," 11; Herrmann, *The Arming of Europe,* 26; Howard, "Men Against Fire," 57; Samuels, *Command or Control?,* 76.

[51] Cassar, *The Tragedy of Sir John French,* 62; J. M. Winter, *The Experience of World War I,* 13.

[52] Beckett, *The Great War,* 45-6; Samuels, *Command or Control?,* 76.

[53] Terence Zuber, *The Battle of the Frontiers, Ardennes 1914*, (Stroud: The History Press, 2010), 32-36.

[54] T Seki, "The Value of the *Arme Blanche*, with illustrations from the recent Campaign," Trans. F. S. G. Piggott, *Royal United Service Institute Journal*, Vol. 55, part 2 (July-Dec. 1911), 894.

[55] André Corvisier, John Childs, Chris Turner, *A Dictionary of Military History and the Art of War* (Cambridge: Blackwell Publishing, 1994), 353; Beckett, *The Great War,* 45 and 49.

[56] Anon, *Reglement sur le service des armee en campagne*, 1913. 77-8 and 94-5. Cited in: Jonathan M. House, "The Decisive Attack: A New Look at French Infantry Tactics on the Eve of World War I," *Military Affairs*, Vol. 40, No. 4 (Dec., 1976), 165.

[57] Arnold, "French Tactical Doctrine," 63-4.

[58] Bidwell and Graham, *Firepower,* 15 and 19.

[59] Sheffield, *Forgotten Victory,* 93; Stone and Schmidl, *The Boer War,* 123; Tim Travers, *The Killing Ground*, 37; ——, "The Offensive and the Problem of Innovation," 539.

[60] Palazzo, *Seeking Victory,* 10.

[61] Ibid, 11 and 12.

[62] Engen, "Steel against Fire," 12.

[63] Stone and Schmidl, *The Boer War,* 10.

[64] Hutton, *Fixed Bayonets,* 1.

[65] Henry Angelo, *The Reminiscences of Henry Angelo* (Manchester: Ayer Publishing, 1972), xvii.

[66] Richard F. Burton, *A Complete System of Bayonet Exercise* (William Clowes and Sons, 1953), 5.

[67] Ibid, 9.

[68] Ibid, 9.

[69] James Dunbar Campbell, "The Army isn't All Work: Physical Culture and the Evolution of the British Army 1860-1918" (PhD thesis: University of Maine, 2003); Clifford, *The British Army,* 59.

[70] Adjutant General's Office, *Bayonet exercise* (London: HMSO, 1860); War Office. *Infantry sword and carbine sword-bayonet exercises* (London: HMSO, 1880).

[71] "A. F. S., "Hutton, Alfred," *Dictionary of National Biography*, ed. Sidney Lee (Adamant Media Corporation, 2001), 332.

[72] Alfred Hutton, *Bayonet Fencing and Sword Practice* (London: W. Clowes and Sons, 1882); ——, *Fixed Bayonets*, (London: W. Clowes, 1890).

[73] Hutton, *Fixed Bayonets,* 1.

[74] Hutton, *Bayonet Fencing and Sword Practice*, 2.

[75] *Infantry Training 1902*, part 1 subsection 89-100.

[76] This pamphlet was republished in 1907 and included in the 1908 reprinting of *Infantry Training 1905*. War Office, *Instruction in Bayonet Fighting* (London: Harrison and Sons, 1907), *Infantry Training 1905*, appendix 2.

[77] *Infantry Training 1902*, part 1 subsection 94.

[78] *Infantry Training 1905*, appendix 2, subsections 2 and 16.

[79] Ibid, appendix 2, subsection 5.

[80] Ibid, appendix 1, section 3.

[81] Ibid, appendix 2, subsection 20.

[82] Ibid, appendix 2, subsection 20.

[83] *Infantry Training 1911*, appendix 1, section 2.

[84] Ibid, appendix 1, section 1, para. 2.

[85] Ibid, part 4, section 125, para. 6.

[86] *Infantry Training 1905*, appendix 2, subsection 20.

[87] *Infantry Training 1911*, appendix 1, section 2.

[88] Ibid, appendix 1, section 2.

[89] Ibid, appendix 1, section 2.

[90] Travers, "The Offensive and the Problem of Innovation," 543.

Chapter III: Fear and Function

[1] Sydney Anglo, *The Martial Arts of Renaissance Europe* (New Haven: Yale University Press, 2000), 37.

[2] Hutton, *Cold Steel* (London: William Clowes and Sons, 1889); ——, *Old Sword Play: the systems of fence in vogue during the XVIth, XVIIth, and XVIIIth centuries with lessons arranged from the works of various ancient masters* (London: Westermann, 1892); ——, *The Sword and the Centuries: or, Old sword days and old sword ways; being a description of the various swords used in civilized Europe during the last five centuries, and of single combats which have been fought with them / With introductory remarks by Cyril G. R. Matthey* (New York: Grant Richards, 1901); Edgerton Castle, *Schools and Masters of the Fence: From the Middle Ages to the Eighteenth Century* (London: Arms and Armour Press, 1969), iii.

[3] Cyril Matthey, *The Complete Works of George Silver, Comprising "Paradoxes of Defence" [Printed in 1599 and now reprinted] and "Bref Instructions Upon my Paradoxes of Defence" [Printed for the first time from the MS. In the British Museum]* (London: George Bell and Sons, 1898), v-xix.

[4] Will R. Bird, *Ghosts Have Warm Hands* (Vancouver: Clarke, Irwin & company Ltd., 1968), 91-2.

[5] *Bayonet Training 1916*, 1 para. 1.

[6] Headquarters Canadian Overseas Military Forces, *Bayonet Fighting Illustrated 1917* (London: Harrison and Sons, 1917), 94

[7] J. Christoph Amberger, *The Secret History of the Sword: Adventures in Ancient Martial Arts* (Burbank: Hammerterz Forum, 1996), 69-86; Bourke, 159.

[8] *Canada in the Great World War... Volume VI*, 281

[9] Thomas Dinesen, *Merry Hell! A Dane with the Canadians* (London: Jarrolds, 1930), 239. Other examples of disassociation see LAC RG 41, vol. 10, 18th Btn., Sid Smith, 1/13

[10] LAC RG 41, vol. 14, 46th Btn., Crowe 2/10.

[11] *Bayonet Training 1916*, 11, para. 32.

[12] For the psychological effect of German fencing with sharpened weapons see: Amberger, *The Secret History of the Sword* 47-53.

[13] Morton, *When your Numbers Up*, 284, note 58; Denis Winter, *Death's Men*, 40.

[14] Hodges, "'They don't like it up 'em'," 135.

[15] Ibid, 124; Bourke, *An Intimate History of Killing*, 54.

[16] James C. Morrison, *Hell on Earth, A personal account of Prince Edward Island soldiers in The Great War* (Summerside: J. C. Morrison 1995), 120.

[17] Daniel Heidt, "From Bayonets to Stilettos to UN Resolutions: The Development of Howard Green's Views Regarding War" (M.A. Thesis, University of Waterloo, 2008), 8.

[18] Hodges, "'They don't like it up 'em'," 129.

[19] *The Organization of the Bayonet Fighting*, 16-7.

[20] Ibid, 17.

[21] Bourke, *An Intimate History of Killing*, 85.

[22] Ibid, 54.

[23] Dinesen, *Merry Hell!*, 43.

[24] Enos Grant, 8 May 1915, From Canadian Letters & Images Project.

[25] *Bayonet Fighting Illustrated 1917*, 5; similar statements about "the spirit of the bayonet" also appear in: *Organization of the Bayonet Fighting*, 7 and *Bayonet Training 1916*, 1.

[26] War Office, *S.S. 143 Instructions for the Training of Platoon for Offensive action* (Milton Keynes, Military Press, 2000), 97.

[27] Matheson, *My Grandfather's War*, 64.

[28] Ibid, 64.

[29] Hodges, "'They don't like it up 'em'," 128.

[30] Taylor, *Illustrated History*, 100 and Denis Winter, *Death's Men*, 110.

[31] Cook, "The Politics of Surrender," 639-40, and 645. Ferguson, "Prisoner Taking and Prisoner Killing 153-5. Niall Ferguson, "Prisoner Taking and Prisoner Killing in the Age of Total War: Toward a Political Economy of Military Defeat," *War in History* 11, No. 2 (April 2004), 149.

[32] D. J. Corrigall, *The History of the Twentieth Canadian Battalion (Central Ontario Regiment) Canadian Expeditionary Force in the Great War, 1914-1918* (Toronto: Stone & Cox, 1935), 162; A. Fortescue Duguid, *History of the Canadian Grenadier Guards: 1760-1964* (Montreal: Gazette Printing, 1965), 112, 194, 207; Featherstonhaugh, *The 13th Battalion*, 197, 252, and 253. ——, *14th Battalion*, 137; D. J. Goodspeed, *Battle Royal: A History of the Royal Regiment of Canada 1862-1962* (Toronto: Charters Publishing, 1962), 205-6; G. Chalmers Johnson, *The 2nd Canadian Mounted Rifles [British Columbia Horse] In France and Flanders* (Vernon: CEF Books, 2003), 70; S. Douglas MacGowan, Harry M. Heckbert, and Byron E. O'Leary, *New Brunswick's 'Fighting 26th': A History of the 26th New Brunswick Battalion, C.E.F. 1914-1919* (Sackville: Neptune Publishing, 1994), 119, 247; McClintock, *Best O' Luck,* 61; Bernard McEvoy and A. H. Finlay, *History of the 72nd Canadian Infantry Battalion: Seaforth Highlanders of Canada* (Vancouver: Cowan and Brookhouse, 1920), 79, 144, and 148; James L. McWilliams and R. James Steel, *The Suicide Battalion* (St. Catharines: Vanwell Publishing, 1990), 197; W. C. Millar, *From Thunder Bay Through Ypres with the Fighting 52nd* (Canada: s.n., 1918), 20; W. W. Murray, *The History of the 2nd Canadian Battalion (East. Ontario Regiment) Canadian Expeditionary Force in the Great War 1914-1919* (Ottawa: Mortimer, 1947),127; H. C. Singer, *History of the 31st Canadian Infantry Battalion C.E.F.* (Calgary: Detselig, 2006), 162, 268, 269,

333, and 373-4; Bruce Tascona and Eric Wells, *Little Black Devils: A History of the Royal Winnipeg Rifles* (Frye Publishing, 1983), 111-2; C. Beresford Topp, *The 42nd Battalion, C.E.F. Royal Highlanders of Canada in The Great War* (Montreal: Gazette Printing, 1931), 211-2; LAC RG 9 III-C-1, vol. 3859, Folder 86, file 13, "Reports on Canadian Corps Operations – Attacks by formations and units 8th July 1917 to 31st December, 1917," 1; LAC RG 9 III-C-3, vol. 4106, Folder 22, file 14, "Operations Minor 6th Canadian Infantry Brigade," "Minor Operation by 31st (Alberta) Bn. Neuville Vitasse night 21st/22nd May 1918," 5; LAC RG 9 III-C-3, vol. 4014, Folder 25, File 2 "Operations: Hill 70. (1st Cdn. Div. 1-7-17 to 25-9-17," "Report on the Capture of Hill 70 and Puits 14 Bis by 1st Canadian Division. 15th Aug. 1917;" LAC RG 9 III-C-3, vol. 4015, Folder 30, file 6, "Operations Amiens 2nd Bde" "2nd Canadian Infantry Brigade narrative of the operations east of Amiens August 8/9 1918, 20; LAC RG 9 III-C-3, vol. 4016 Folder 31, File 2 "2nd Canadian Infantry Brigade Narrative of Operations East of Arras August 25th to September 3rd, 1918;" LAC RG 9 III-C-3, vol. 4017, Folder 34, file 9, "Operations Minor (10th Bn) Narrative of Minor Enterprise Carried out on the night of 4/5 February 1916;" LAC RG 9 III-C-3, vol. 4132 Folder 12, File 8 Headquarters, 6th Canadian Infantry Brigade, Narrative Report of Operations for Capture of Passchendaele, Nov 6th, 1917," 5 and 7; Two instances in LAC RG 9 III-C-3, vol. 4229 Folder 19, File 5, Operation Avion 12th CIB S.G. 23/866, "Report on Operations carried out by the 12th Canadian Infantry Brigade between 26th June and 28th June 1917;" LAC RG 9, III-D-3, vol. 4925, WD, 16th Btn., 22 April 1915; LAC RG 9, III-D-3, vol. 4930, WD, 20th Btn., 15 August 1917; LAC RG 9, III-D-3, vol. 4937, WD, 31st Btn., November 1917, Appendix B-1, "Narrative of Operations-PASSCHENDAELE ATTACK, November 5/6/7, 1917," 1; LAC RG 9, III-D-3, vol. 4873, WD, 2nd Bde., August 1918, Appendix 15, 20; LAC RG 41, vol. 11, 27th Btn., John Nind 2/1; LAC RG 41, vol. 15, 54th Btn., A. W. Sturdwick, 1/17.

[33] LAC RG 9 III-C-3, vol. 4014, Folder 25, File 2 "Operations: Hill 70. (1st Cdn. Div). 1-7-17 to 25-9-17."

[34] Topp, *The 42nd Battalion*, 87.

[35] Cook, "The Politics of Surrender," 646 and 649.

[36] Goodspeed, *Battle Royal*, 205-6.

[37] Cook, "The Politics of Surrender," 645.

[38] Original emphasis. LAC RG 9 III, 3870, Folder 112, File 11,

[39] Cook, "The Politics of Surrender," 646.

[40] Jack Sheldon, *The German Army on Vimy Ridge 1914-1917* (Barnsley: Pen & Sword, 2008), 292.

[41] Bird, *Ghosts Have Warm Hands* (Vancouver: Clarke, Irwin & company Ltd., 1968), 48.

[42] LAC RG 9 III-C-3, vol. 4028, Folder 17, file 20: "Operation: Notes on Lessons Learned," Ref. 8-229, 2nd Btn., August 23rd. 1918.

[43] *Letters From the Front*, 20; Beattie, *48th Highlanders of Canada*, 324; Bird, *Ghosts Have Warm Hands*, 203; Fraser, *The Journal of Private Fraser: 1914-1918 Canadian Expeditionary Force*, Ed. Reginald H. Roy (Victoria: Sono Nis Press, 1985), 89; Joseph Hayes, *The Eighty-Fifth in France and Flanders*, 149; MacGowan, Heckbert, and O'Leary, *New Brunswick's 'Fighting 26th'*, 247; McEvoy and Finlay, *History of the 72nd Canadian Infantry Battalion*, 72; W. W. Murray, *The History of the 2nd Canadian Battalion*, 126; Topp, *The 42nd Battalion*, 79 and 224; LAC RG 9 III-C-3, vol 4091, Folder 26, File 20, "Detailed Report of O.C. 29th (Vancouver) Battalion Report on Minor operation Jan 30th-31st, 1916," 3; LAC RG 9 III-C-3, vol. 4106, Folder 22, file 14 Operations Minor (6th CIB "Narrative of Minor operation Carried out on night June 2nd/3rd opposite front of Mercatel sector by 6th Canadian Infantry Brigade--29th (Vancouver) Bn. June 2/3. 1918; LAC RG 9 III-C-3, vol. 4229, Folder 19, File 5, Operation Avion 12th CIB S.G. 23/866 Report on Operations carried out by the 12th Canadian Infantry Brigade between 26th June and 28th June 1917; LAC RG 9, III-D-3, vol. 4914, WD, 3rd Btn., 13 June 1916; LAC RG 9, III-D-3, vol. 4925, WD, 16th Btn., 22 April 1915; LAC RG 9, III-D-3, vol. 4944, WD, 85th Btn., April 1917, Appendix A "covering operations 8-4-17 to 14-4-17, inclusive," 3; LAC RG 9, III-D-3, vol. 4944, WD, 85th Btn., October 1917, Appendix A, "Report on move of Battalion into the line and report on operation." 3; LAC RG 41, vol. 12, 28th Btn., Arthur B. Goodmurphy, 2/17; LAC RG 41, vol. 7, 2nd Btn. W. F. Graham, 1/8; LAC RG 41, vol. 15, 50th Btn., Mr. A. Turner, 3/6; LAC RG 41 vol. 18 Fort Gary Horse Lt. Col. Strachan 1/9.

[44] Cook, "The Politics of Surrender," 661-2, and 664.

[45] Timothy T. Lupfer, *Leavenworth Papers, No. 4, Dynamics of Doctrine: The Changes in German Tactical Doctrine During the First World War* (Leavenworth: U.S. Army Command and General Staff College, 1981), 15-6; Samuels, *Command or Control?*, 173, 175, 186, 192, and 196.

[46] Bourke, *An Intimate History of Killing*, 252.

[47] H. M. Urquhart, *The History of the 16th Battalion (The Canadian Scottish) Canadian Expeditionary Force in the Great War, 1914-1919*, (Toronto: MacMillan, 1932), 144.

[48] Bird, *Ghosts Have Warm Hands*, 174.

[49] Original emphasis. LAC RG 9, III, 3870, Folder 112, File 11.

⁵⁰ War Office. *Bayonet Fighting: Instruction with Service Rifle and Bayonet: 1915* (London: Harrison and Sons, 1915), 3; 13, para. 41; Other references to the yell in training see: Louis Keene, *Crumps: The Plain Story of a Canadian who went* (Boston: Houghton Mifflin, 1917), 53; E. S. Russenholt, *Six Thousand Canadian Men: Being the History of the 44th Battalion Canadian Infantry 1914-1919* (Winnipeg: De Montford Press, 1932), 19. LAC RG 9 III, vol. 3870, Folder 112 File 11, "Bayonet Fighting for Platoon Commanders;" "INDIVIDUAL ASSAULT PRACTICE;"

⁵¹ *Letters From the Front*, 17; Featherstonhaugh, *The 13th Battalion*, 197; Joseph Hayes, *The Eighty-Fifth in France and Flanders*, 91; McEvoy and Finlay, *History of the 72nd Canadian Infantry Battalion*, 144; G. B. McKean. *Scouting Thrills* (New York: MacMillan Company, 1919); 211-2; McWilliams and Steel, *The Suicide Battalion*, 164; Tascona and Wells, *Little Black Devils*, 112; Topp, *The 42nd Battalion*, 210; Urquhart, *The History of the 16th Battalion*, 59; LAC RG 9 III-C-3, vol. 4011, Folder 17, File 4 "Operations: Somme, 1916 (3rd Canadian Infantry Bde). "Finding of the Court of Enquiry held at Bouzincourt on the evening of February, 1916, to investigate and enquire into certain matters pertaining to the operations of the 1st Canadian Brigade during the operations of the 8th October, 1916," 2; LAC RG 9 III-C-3, vol. 4132, 6th Brigade, "Narrative of Offensive Operations on April 9th and 10th, 1917," 3; LAC RG 9, III-D-3, vol. 4935, WD, 27th Btn., April 1917, Appendix L "Narrative of Offensive 9.4.17," 2; LAC RG 41, vol. 7, 2nd Btn., W. F. Graham Tape 1/5; LAC RG 41, vol. 14, 46th Btn., Bob Bron, 3/1; LAC RG 41, vol. 16, 78th Btn., Oscar Ericson 2/4; LAC RG 41, vol. 18, Fort Gary Horse, Lt. Col. Strachan 1/9.

⁵² Duguid, *Official History, Part I*, 403 and *Part II*, 193.

⁵³ Cook, "The Politics of Surrender," 641.

Chapter IV: 1915, The Bayonet and Trench Warfare

¹ Stevenson, *Cataclysm*, 45; Beckett, *The Great War*, 55.

² Kramer, *The Dynamic of Destruction*, 34-35; Beckett, *The Great War*, 166.

³ Stevenson, *Cataclysm*, 146.

⁴ Cook, *At the Sharp End*, 223; Rawling, *Surviving Trench Warfare*, 42.

⁵ Morton, *When Your Number's Up*, 120.

⁶ Cook, *At The Sharp End*, 65.

[7] Johnathan Bailey, "British Artillery in the Great War," *British Fighting Methods in the Great War* (Portland: Frank Cass, 1996), 23; Bidwell and Graham, *Firepower,* 9-10; Cook, *At The Sharp End,* 175; Morton, *When Your Number's Up,* 150; Sheffield, *Forgotten Victory,* 99.

[8] Sheffield, *Forgotten Victory,* 106.

[9] Gudmundsson, *Stormtroop Tactics,* 29.

[10] Bailey, "British Artillery in the Great War," 154; Duguid, *Official History,* Part I, 176; Rawling, *Surviving Trench Warfare,* 41; Duguid, *Official History,* Part I, 176; Winter and Baggett, *The Great War,* 59-68.

[11] Strachan, *The First World War,* 1060; Rawling, *Surviving Trench Warfare,* 41.

[12] Griffith, *Battle Tactics,* 51; Chris McCarthy, "Queen of the Battlefield: The Development of Command, Organisation and Tactics in the British Infantry Battalion during the Great War," *Command and Control on the Western Front: The British Army's Experience, 1914-1918,* ed. Gary Sheiffeild and Dan Todman (Staplehurst: Spellmount, 2004), 176.

[13] Duguid, *Official History, Part I,* 175.

[14] Ramsay, *Command and Cohesion,* 170; David Charles Gregory Campbell, "The Divisional Experience in the C.E.F.: A Social and Operational History of the 2nd Canadian Division, 1915-1918." (PhD thesis: University of Calgary, 2003), 225.

[15] Nicholson, *C.E.F.*, 20 and 24; Duguid, *Official History,* 66-67.

[16] This story in found in Ronald G. Haycock, *Sam Hughes: The Public Career of a Controversial Canadian 1885-1916* (Waterloo: Wilfrid Laurier Press, 1986), 143; Cook, *At the Sharp End,* 47; Morton, *When Your Number's Up,* 16. There is no mention of bayonet training in the daily orders for Valcartier in August and September 1914, (LAC RG 9 II, F-9, vol. 1702). Nor is there mention of bayonet training in the War Diaries of the 1st and 4th Battalions, which began keeping records in September 1914 (see LAC RG 9, Series III-D-3, vol. 4912, Reel T-10704 and vol. 4915, Reel T-10707). The remainder of the Battalion diaries do not begin record keeping before October 1914. Sam Hughes was more obsessed with musketry in 1914 and spoke derisively of the bayonet in deliberations on the development of the Ross Rifle Mark II in 1904-5. "He 'had little use for a bayonet, that is for practical purposes, but it serves a useful sentimental object.'" (Duguid, *Official History,* Part II, 80.).

[17] Duguid. *Official History,* Part I, 131.

[18] *Infantry Training 1914,* Appendix 2, section 1.

[19] Ramsay, *Command and Cohesion,* 162; Bidwell and Graham, *Firepower,* 61.

[20] Duguid, *Official History*, Part 1, 131.

[21] Iarocci, *Shoestring Soldiers*, 72.

[22] Duguid, *Official History*, Part II, Appendix 276, 198.

[23] Ibid, Appendix 276, 198.

[24] *Memorandum of the Training and Employment of Grenadiers,* cited in Duguid, *Official History,* Part II, Appendix 276, 199.

[25] *Letters from the Front*, 22.

[26] Duguid, *Official History,* Part I, 175.

[27] LAC RG 9 III-C-3, vol. 4079, Folder 8, file 6, 15th Battalion Training.

[28] Russenholt, *Six Thousand Canadian Men,* 167.

[29] For other references to bayonet fighting at the 2nd Battle of Ypres see: *Letters from the Front,* 16; Duguid, *Official History,* Part I, 293; Iarocci, *Shoestring Soldiers*, 149; Nicholson, *C.E.F.*, 67; W. W. Murray, *The History of the 2nd Canadian Battalion,* 49; Gordon Reid, *Poor Bloody Murder: Personal Memoirs of the First World War*, 67; LAC RG 41, vol. 9, 16th Btn., H. H. Oldaker, 2/6.

[30] Iarocci, *Shoestring Soldiers*, 113.

[31] LAC RG 41, vol. 9, 16th Btn., Mr. B. C. Lunn 2/4

[32] Urquhart, *The History of the 16th Battalion,* 59.

[33] LAC RG 9, III-D-3, vol. 4919, WD, 10th Btn., 22 April 1915.

[34] LAC RG 41, vol. 8, 10th Btn., Sid Cox. 1/4,

[35] Duguid, *Official History,* Part I, 243.

[36] *Letters from the Front*, 15.

[37] LAC RG 9, III-D-3, vol. 4919, WD, 10th Btn., 22 April 1915; LAC RG 9, III-D-3, vol. 4874, WD 3rd Bde., May 1915, appendix A, "Diary of operations, 3rd Canadian Infantry Brigade, 22nd April to 5th May, 1915," 2,

[38] LAC RG 41, vol. 9, 16th Btn., Mr. B. C. Lunn 2/6.

[39] LAC RG 9, III-D-3, vol. 4925, WD, 16th Btn., 22 April 1915.

[40] Cook, *At the Sharp End*, 125 and 129; Iarocci, *Shoestring Soldiers*, 117; Duguid, *Official History*, Part I, 244-5.

[41] Andrew Iarocci, "1st Canadian Infantry Brigade in the Second Battle of Ypres, The Case of the 1st and 4th Canadian Infantry Battalions, 23 April 1915," *Canadian Military History* 12, No. 4 (Autumn 2003), 8-11.

[42] Iarocci, *Shoestring Soldiers*, 126-134; ——, 11-12; "1st Canadian Infantry Brigade;" LAC RG 9, III-D-3, vol. 4912, WD, 1st Btn., April 1915, Appendix 1, "Narrative of operations 23rd to 30th April, 1915;" LAC RG 9, III-D-3, vol. 4916, WD, 4th Btn., April 23rd, 1915.

[43] Additional references to bayonet fighting and hand-to-hand fighting at Festubert and Givenchy not investigated in this chapter, see: *Letters from the Front*, 17; Alexander Ewen, 29 May 1915, Canadian Letters & Images Project,

[44] Cook, *At the Sharp End*, 177; Duguid, *Official History*, Part I, 464.

[45] Iarocci, *Shoestring Soldiers*, 199.

[46] Ibid, 207; LAC RG 9, III-D-3, vol. 4919, WD, 10th Btn., 20 May 1915.

[47] Beattie, *48th Highlanders of Canada,* 88

[48] LAC RG 9, III-D-3, vol. 4924, WD, 15th Btn., 20 May 1915.

[49] Rawling, *Surviving Trench Warfare,* 43.

[50] Iarocci, *Shoestring Soldiers*, 209.

[51] P. A. Guthrie, "Festubert: A Graphic Story of a Great Fight Where Canadians Won Honor at a Heavy Price Just a Year Ago," *The Montreal Daily Star*, 27 May 1916, 17. Percy Guthrie was the commanding officer of the 10th Bn and Festubert.

[52] Sic. LAC RG 41, vol. 9, 16th Btn., A.M. McLennan, 2/8.

[53] Iarocci, *Shoestring Soldiers*, 210.

[54] Guthrie, "Festubert," 17.

[55] LAC RG 9, III-D-3, vol. 4919, WD, 10th Btn., 21 May 1915.

56 Guthrie, "Festubert," 17.

57 Guthrie, "Festubert," 17.

58 LAC RG 9, III-D-3, vol. 4916, WD, 5th Btn., May 1915, Appendix A, "Festubert."

59 Iarocci, *Shoestring Soldiers*, 213.

60 *Letters From the Front*, 20.

61 Baldwin, *Holding the Line*, 263-4.

62 LAC RG 9, III-D-3, vol. 4914, WD, 3rd Btn., May 1915, Appendix 1, "Operation Orders by Lieut. Col. Rennie," 1.

63 Ibid, 1.

64 Duguid, *Part I*, 488-9.

65 Iarocci, *Shoestring Soldiers*, 229; LAC RG 9, III-D-3, vol. 4867, WD, 1st Bde., 15 June 1916, 11:20 p.m.

66 Iarocci, *Shoestring Soldiers*, 229.

67 LAC RG 9, III-D-3, vol. 4912, WD, 1st Btn., June 1915, appendix 2, "Operation Order No. 5," 2

68 LAC RG 9, III-D-3, vol. 4867, WD, 1st Bde., 15 June 1915, 11:20 p.m.

69 LAC RG 9, III-D-3, vol. 4912, WD, 1st Btn., June 1915, appendix 1, "Narrative of Operations, 15th June, 1915," 2

70 Rawling, *Surviving Trench Warfare*, 79.

71 LAC RG 9 III-C-3, vol. 4217, Folder 7, file 19, "Training: vol. 1, 50th Battalion."

72 An Officer, *Practical Bayonet Fighting: With Service Rifle and Bayonet* (London: The Bazaar, Exchange and Mart Office, Windsor House, Bream's Buildings, E.C., 1915); Leopold McClaglen, *The McClagken System of Bayonet Fighting*, (Christchurch: Christchurch Press, 1915); ——- *Bayonet Fighting for War* (London: Harrison and Sons, 1916). Several titles of other unofficial Bayonet Manuals are to be found in *The English Catalogue of Books for 1915*, (London: Publishers Circular, 1916): Anon, *Cold Steel : How to Use the Bayonet, Sword and Lance. With a chapter on the Pistol at close quarters. Written and illustrated by the Staff*

of The Regiment. (Temple Press, 1915). Cornish, *Bayonet, &c., How to use, Cold steel,* (1915); and Wyman, *Army Bayonet fighting: Instruction with service rifle and bayonet,* (1915).

[73] LAC RG 9 III-C-3, vol. 4217, Folder 7, file 19, "Training: vol. 1, 50th Battalion."

[74] Machum, *The Story of the 64th Battalion,* 58-60.

[75] McWilliams and Steel, *The Suicide Battalion,* 24.

[76] Urquhart, *The History of the 16th Battalion,* 119.

[77] David Charles Gregory Campbell, "The Divisional Experience in the C.E.F.," 153.

[78] LAC RG 41, vol. 8, 5th Btn., F. G. Baghaw, 2/11.

[79] sic. Baldwin, *Holding the Line,* 97-99.

[80] George Yale Harrison, *Generals Die in Bed* (Toronto: Annick Press, 2004), 78. John William Lynch, *Princess Patricia's Canadian Light Infantry, 1917-1919* (Hicksville: Exposition Press, 1976), 126-7; Reid, *Poor Bloody Murder,* 107; LAC RG 41, vol. 8, 5th Btn., J. Younie, 1/5; LAC RG 41, vol. 17, 5th CMR, Henry Newmark, 2/13.

[81] *Bayonet Fighting Illustrated 1917,* 38; *Bayonet Training 1916,* para. 23.

[82] An Officer, *Practical Bayonet Fighting,* 33.

[83] McMillan, *Trench Tea and Sandbags,* 6.

[84] ibid. 6.

[85] An Officer, *Practical Bayonet Fighting,* 14.

[86] LAC RG 24, vol. 1862, "Cummins Notes on the First World War," File 102 "Physical Training and Bayonet Fighting"

[87] Duguid, *Official History,* Part I, 426; LAC RG 9 III, vol. 678, file H 25-2-86, Correspondence from Mayes to Col. Reid dated 20 November 1915.

[88] LAC RG 9 III, vol. 678, file H 25-2-86, Correspondence from Mayes to Col. Reid dated 20 November 1915.

[89] LAC RG 24, vol. 1862. Cummins Notes on the First World War File 102 Physical Training and Bayonet Fighting

[90] Original boldface, *Bayonet Fighting 1916*, para.13.

[91] *Bayonet Fighting 1915*, 3.

Chapter V: 1916, The Bayonet and the Battle of Materiel

[1] Cook, *At the Sharp End*, 473.

[2] Original emphasis, *Letters From the Front*, 105.

[3] Mark Osborne Humphries, "Old Wine in New Bottles," 66; Rawling, *Surviving Trench Warfare*, 71; Tim Travers, "Learning and Decision-Making," 93.

[4] Lupfer, *Dynamics of Doctrine*, 7.

[5] Paddy Griffith, "The Extent of Tactical Reform in the British Army," *British Fighting Methods*, 11-12; Morton, *When Your Number's Up*, 131.

[6] Bailey, "British Artillery in the Great War," 28; Beckett, *The Great War*, 167; Cook, *At the Sharp End*, 410; Morton, *When Your Number's Up*, 154; Palazzo, *Seeking Victory*, 92-3; Samuels, *Command or Control?*, 131.

[7] Jack Sheldon, *The German Army on the Somme 1914-1916* (Barnsley: Sword and Pen, 2006), 354, note 50.

[8] Griffith, "The Extent of Tactical Reform," 12; Samuels, *Command or Control?*, 136; Rawling, *Surviving Trench Warfare*, 71.

[9] David Charles Gregory Campbell, "The Divisional Experience in the C.E.F.," 140-2.

[10] Cook, "The Blind leading the Blind," 34; Rawling, *Surviving Trench Warfare*, 61.

[11] Cook, "The Blind leading the Blind," 31.

[12] Cook, *At the Sharp End*, 350.

[13] Ralph Hodder-Williams, *Princess Patricia's Canadian Light Infantry, 1914-1919*, (Hodder and Stroughton, 1923), 126; Nicholson, *C.E.F.*, 67.

[14] Cook, *At the Sharp End*, 351; Nicholson, *C.E.F.*, 134.

[15] Cook, *At the Sharp End*, 358; Johnson, 24; Nicholson, *C.E.F.*, 134.

[16] Stewart, 49.

[17] Morton, *When Your Number's Up*, 153.

[18] Urquhart, *The History of the 16th Battalion,* 144; LAC RG 9, III-D-3, vol. 4912, WD, 1st Btn., June 1916, Appendix 1, "Report on operations carried out by 1st Canadian Infantry Battalion on the occasion of an attack on the German position at Mount Sorrel near Ypres June 12th and 13th, 1916."

[19] Cook, *At the Sharp end*, 372.

[20] Urquhart, *The History of the 16th Battalion,* 144.

[21] LAC RG 9, III-D-3, vol. 4914, WD, 3rd Btn., 13 June 1916.

[22] Ibid.

[23] LAC RG 41, vol. 7, 3rd Btn., Alley, 2/4.

[24] Goodspeed, *Battle Royal*, 145; Urquhart, *The History of the 16th Battalion,* 144; Featherstonhaugh, *The 13th Battalion*, 103; LAC RG 9, III-D-3, vol. 4921, WD, 13th Btn., 12/13 June 1916.

[25] LAC RG 9, III-D-3, vol. 4925, WD, 16th Btn., 13 June 1916.

[26] LAC RG 9, III-D-3, vol. 4914, WD, 3rd Btn., 13 June 1916.

[27] Goodspeed, *Battle Royal*, 146.

[28] Urquhart, *The History of the 16th Battalion,* 144.

[29] LAC RG 9, III-D-3, vol. 4925, WD, 16th Btn., 13 June 1916.

[30] LAC RG 9, III-D-3, vol. 4914, WD, 3rd Btn., 13 June 1916.

[31] Other records of bayonet and hand-to-hand fighting at Mount Sorrel see: Cook, *At the Sharp End*, 351; Featherstonhaugh, *The 13th Battalion*, 103-4; Stuart Martin, *The Story of the 13th Battalion, 1914-1917* (London: Canadian War Records Office, 1917), 11; Nicholson, *C.E.F.*, 134.

[32] Stewart, "Attack Doctrine in the Canadian Corps," 13; Todman and Sheffield, 8,

[33] Bidwell and Graham, *Firepower*, 2; Lupfer, *Dynamics of Doctrine*, 7; Rawling, *Surviving Trench Warfare*, 68-9; Samuels, *Command or Control?*, 140-142; Terraine, *The Great War*, 253.

[34] Herrmann, *The Arming of Europe*, 223; Ramsay, *Command and Cohesion*, 195.

[35] Bailey, "British Artillery in the Great War," 32.

[36] Christopher Duffy, *Through German Eyes, The British and the Somme 1916* (London: Weidenfeld & Nicolson, 2006), 288; Nicholson, *The Gunners of Canada*, 272-3; Rawling, *Surviving Trench Warfare*, 69.

[37] RG 9, III, C. 1 vol 3859 Folder 86, file 6, "Notes on the Somme Fighting"; RG 9, III-C-3, vol. 4011, Folder 17, File 1 "Operations Somme 1916: Notes on, experienced gained, lessons learned"; RG 9 III-C-3 vol 4089, Folder 20, file 4 Operations, Somme 1916 2-8-16 to 25-11-16, 2nd Division notes on the Somme Battle," "A Notes on 1st Phase"; RG 9 III-C-3, vol 4139, "Operations: Somme, 28th Battalion," "Following Note drawn from the fighting on the Somme."

[38] Rawling, *Surviving Trench Warfare*, 69-70.

[39] Bailey, "British Artillery in the Great War," 29; Radley, *We Lead Others Follow*, 90.

[40] Bidwell and Graham, *Firepower*, 83-5.

[41] Duffy, *Through German Eyes*, 142; Griffith, "The Extent of Tactical Reform," 10; ---, *Battle Tactics*, 65.

[42] Duffy, *Through German Eyes*, 150; Rawling, *Surviving Trench Warfare*, 84.

[43] Original emphasis. LAC RG 9 III, C. 1, vol. 3859, Folder 86, file 6, "Notes on the Somme Fighting" (25th Division), 3.

[44] Other accounts of bayonet fighting in the second phase of the attack at Courcelette see: Calder, *The History of the 28th (Northwest) Battalion*, 90; Topp, *The 42nd Battalion*, 79 and 87.

[45] *Letters From the Front*, 186.

[46] Featherstonhaugh, *The 24th Battalion*, 90; LAC RG 9, III-D-3, vol. 4930, WD, 21st Btn., 15 September 1916; LAC RG 9, III-D-3, vol. 4884, WD, 5th Bde., September 1916, appendix f, "Summary of Operations from 11th to 17th September, 1916," 4.

47 Cook, *At the Sharp End*, 461.

48 D.G. Scott Calder, *The History of the 28th (Northwest) Battalion, C.E.F. (October 1914-june 1919)* (Regina: Regina Rifle Regiment, 1961), 28.

49 LAC RG 41, vol. 10, 18th Btn., Sid Smith, 1/13.

50 LAC RG 41, vol. 12, 28th Btn., Arthur B. Goodmurphy, 1/5.

51 Nicholson, *C.E.F.*, 161-2.

52 MG 30 E297, vol. 1, Frank Maheux papers, letter 20th September 1916. Citted in Cook, *At the sharp end*, 450.

53 LAC RG 9, III-D-3, vol. 4895, WD, 8th Bde., September 1916, Appendix f, "Summary of Operations from Sept. 11th to 17th," 2-3.

54 Ibid, 5.

55 Ibid, 5.

56 Ibid, 5.

57 Topp, *The 42nd Battalion*, 79; MacGowan, Heckbert, and O'Leary, *New Brunswick's 'Fighting 26th'*, 118 and 119.

58 LAC RG 9 III-C-3 vol. 4089, Folder 20, file 11, "Operations: Courcelette 8-9-16 to 17-11-16," "2nd Division HQ, 5th Canadian Infantry Brigade," "Formations adopted by the Battalions of the 5th Canadian Infantry Brigade in the ATTACK ON COURCELETTE on the afternoon of Sept. 15th '16." 2.

59 Ibid, 1.

60 Joseph Chabelle, cited in Cook, *At the sharp end*, 460-1

61 Additional references to bayonet fighting occurring on the Somme not examined in this chapter see: Duguid, *Canadian Grenadier Guards*, 112; Featheringstonhaugh, *The 24th Battalion*, 90; Johnson, *The 2nd Canadian Mounted Rifles*, 33; Martin, *The Story of the 13th Battalion*, 13; McClintock, *Best O' Luck*, 57, 60-1; McWilliams and Steel, *The Suicide Battalion*, 57; George Stirrett, *A Soldiers Story – 1914-1918* (Unpublished manuscript, Canadian War Museum), 11; Urquhart, *The History of the 16th Battalion*, 166; *Canada in the Great World War, Volume. VI*, 276; LAC RG 9, III-D-3, vol. 4939, WD, 46th Btn., November 1916, Appendix 3, "Report on Operations night of November 10/11th, 1916," 1; LAC RG 41, vol. 14, 49th Btn., C. P Keeler, 1/13; LAC RG 41, vol. 17, 2nd CMR, S. Bowe,

2/2; LAC RG 41, vol. 17, 116th Btn., General Pearkes, 2/5.

[62] Cook, *At the Sharp End*, 490-3.

[63] Nicholson, *C.E.F.*, 162; WD, 4th CMR, October 1916, Appendix 1, "Narrative of events during the action of the 1st October, 1916," 2; WD, 5th CMR, 1 October 1916; WD, 25th Btn., October 1916, Appendix A, Untitled report 1 October 1916 Operations, 1.

[64] Cook, *At the Sharp End*, 493; Nicholson, *The Gunners of Canada*, 267; Rawling, 76; WD, 24h Btn., Appendix A, "Summary of Operations covering tour, September 27th to October 2nd, 1916," 3.

[65] WD, 5th CMR, 1 October 1916.

[66] ibid.

[67] RG 41, vol. 17, 116th Btn., General Pearkes, 2/5.

[68] The official history also notes hand-to-hand fighting. Nicholson, *C.E.F.*, 162.

[69] Rawling, *Surviving Trench Warfare*, 76.

[70] Urquhart, 182-3.

[71] Urquhart, *The History of the 16th Battalion*, 183.

[72] WD, RCR, October 1916, Appendix 3 & 4, "Narrative of operations of the Royal Canadian Regiment on 7th, 8th, and 9th October, 1916," 2.

[73] Ibid, 2.

[74] Ibid, 2.

[75] LAC RG 41, vol. 18, RCR, H. Arnson Green, 1/8

[76] LAC RG 9, III-D-3, vol. 4911, WD, RCR, October 1916, Appendix 3 & 4, "Narrative of operations of the Royal Canadian Regiment on 7th, 8th, and 9th October, 1916," 3-4.

[77] Goodspeed, *Battle Royal*, 164.

[78] WD, 3rd Btn., 8 Oct. 1916.

[79] Ibid.

[80] RG 9 III-C-3, vol. 4011, Folder 17, File 4, "Operations: Somme, 1916 (3rd Canadian Infantry Bde)." "Finding of the Court of Enquiry held at Bouzincourt on the evening of February, 1917, to investigate and enquire into certain matters pertaining to the operations of the 1st Canadian Brigade during the operations of the 8th October, 1916," 3.

[81] Ibid., 3.

[82] Nicholson, *The Gunners of Canada*, 267; LAC RG 9, III-D-3, vol. 4940, WD, 49th Btn., October 1916, "Report on the operation of the Bn. between the 7th and 9th," 5-6.

[83] LAC, RG 9, III, 678. H 25-2-86, 10th Jan 1916 Correspondence to Henry George Mayes sent by Maj Gen Carson from Shorncliffe. RE: Establishment for the Bayonet Fighting Schools. sent by Maj Gen Carson from Shorncliffe. Establishment for the Bayonet Fighting Schools.

[84] LAC RG 9 III, vol. 678. H 25-2-86, Correspondence dated 11 January 1916 and 26 January 1916, from John W. Carson to Brig-Gen. J. C. MacDougall, CO Canadian Training Division Shorncliffe.

[85] RG 24, vol. 1210, HQ 602-13-26. Correspondence dated 23rd March, 1916.

[86] ACI 1968, "Training of Category A Infantry Recruits," 15 October 1916, From *Army Council Instructions*. Duguid, *History of the Canadian Grenadier Guards*, 130-1.

[87] *The Organization of Bayonet Fighting*, 3.

[88] LAC RG 24, 1862. "Cummins Notes on the First World War, File 102 Physical Training and Bayonet Fighting;"

[89] LAC RG 9 III, vol. 678, E-150-2 Mayes to HQ Canadian Training Division Shorncliffe, 3 May 1916.

[90] *The Organization of Bayonet Fighting*, 2.

[91] Army Council Instruction 393, 20 February 1916. *Army Council Instructions*.

[92] One example is found in LAC RG 9 III-C-3, vol. 4240, Training: 85th Battalion, "Progressive Steps in Bayonet Fighting." Undated, between correspondences dated 16 and 28 August 1916.

[93] Dinesen, *Merry Hell!*, 88.

[94] *The Organization of Bayonet Fighting*, 9.

[95] George Hedley Kempling, 12 July 1916. From Canadian Letters & Images Project.

[96] *Bayonet Training 1916*, para 33.

[97] Ibid, para 34.

Chapter VI: 1917, The Bayonet and the Set-Piece Battle

[1] Bidwell and Graham, *Firepower,* 61; Iarocci "The 1st Canadian Division," 156.

[2] H. M. Urquhart, The History of the 16th Battalion (The Canadian Scottish) Canadian Expeditionary Force in the Great War, 1914-1919, (Toronto: MacMillan, 1932), 190-193.

[3] Cook, *Shock Troops,* 27-8; Mark Osborne Humphries, "Old Wine in New Bottles," 69.

[4] LAC RG 9 vol. 4142, folder 6, file 2, "Notes of French Attacks, North-East of Verdun in October and December 1916," 13. cited in Rawling, *Surviving Trench Warfare,* 91.

[5] Rawling, *Surviving Trench Warfare,* 187.

[6] LAC RG 9 III-C-3, vol. 4044, Folder 4, file 10, Training, 4th Canadian Infantry Battalion, "Southern Command: Notes on Training," Dated: 12th August, 1914. Ramsay, *Command and Cohesion,* 161,

[7] McCarthy, 174.

[8] Rawling, *Surviving Trench Warfare,* 175.

[9] Palazzo, *Seeking Victory,* 172-3.

[10] David Charles Gregory Campbell, "The Divisional Experience in the C.E.F.," 281.

[11] Rawling, *Surviving Trench Warfare,* 116

[12] Morton, *When Your Number's Up,* 168; Nicholson, *The Gunners of Canada,* 282.

[13] Nicholson, *The Gunners of Canada*, 282; Rawling, *Surviving Trench Warfare*, 110.

[14] Cook, "The Politics of Surrender," 641; Rawling, *Surviving Trench Warfare*, 122; Sheldon, *The German Army on Vimy Ridge*, 252, 267 and 271-2.

[15] Urquhart, *The History of the 16th Battalion*, 214.

[16] LAC RG 9, III-D-3, vol. 4923, WD 14th Btn., April 1917, Report on Operations dated 12th April 1917.

[17] Rawling, *Surviving Trench Warfare*, 123.

[18] LAC RG 9, III-D-3, vol. 4923, WD 14th Btn., April 1917, Report on Operations dated 12th April 1917.

[19] Rawling, *Surviving Trench Warfare*, 123. For other accounts of bayonet fighting at Vimy Ridge see: Cook, *Shock Troops*, 109 and 137; Featherstonhaugh, *The 24th Battalion*, 121 and 132; ——, *The Royal Canadian Regiment, 1883-1933* (Montreal: Gazette Publishing, 1936), 278; G. E. Hewitt, *The Story of the 28th Battalion, 1914-1917* (London: Canadian War Records Office, 1917), 18; J. A. Holland, *The Story of the 10th Battalion, 1914-1917* (London: Canadian War Records Office, 1917), 79; Johnson, *The 2nd Canadian Mounted Rifles*, 47; McEvoy and Finlay, *History of the 72nd Canadian Infantry Battalion*, 50; Nicholson, *C.E.F.*, 231, 234, and 240; Sheldon, *The German Army on Vimy Ridge*, 291, 295, 296, 298, and 309; Rawling, *Surviving Trench Warfare*, 124; Urquhart, *The History of the 16th Battalion*, 214; LAC RG 9 III-C-3 vol 4012, Folder 19, file 2 "Report on the Operations Carried out by the 1st Canadian Division April 9th –May 5th 1917," 15; LAC RG 9, III-D-3, vol. 4911, WD, RCR, April 1917, Appendix 2, "Summary of Operations," 2; LAC RG 41, vol. 12, 31st Btn., W. McCombie-Gilbert 2/11-2.

[20] LAC RG 9, III-D-3, vol. 4889, WD, 6th Bde., April 1917, Appendix 10a, "Narrative of Offensive operations on 9th and 10th April, 1917," 3; LAC RG 9, III-D-3, vol. 4935, WD, 27th Btn., April 1917, Appendix L "Narrative of Offensive 9.4.17."

[21] LAC RG 9, III-D-3, vol. 4944, WD, 85th Btn., April 1917, Appendix A, "covering operations 8-4-17 to 14-4-17, inclusive," 3.

[22] Joseph Hayes, *The Eighty-Fifth in France and Flanders*, 59.

[23] For accounts of Bayonet fighting at Arleux, Fresnoy and in the Souchez sector see: Holland, *The Story of the 10th Battalion*, 80; MacEvoy and Finlay, 64 and 72; LAC RG 9 III-C-3 vol. 4012, Folder 19, file 2 "1st Canadian Division report on the Vimy Rdige- Willerval-Arleux And Fresnoy Operations: April 9th-May 5th, 1917, June 1917;" LAC RG 9 III-C-3, vol. 4074, Folder 2, File 7 Operations: Vimy Ridge

to Fresnoy, "1 CIB Report of Recent Operations Carried out be the 1st Canadian Infantry Division, April 9th to may 5th, 1917." part 3, 27; LAC RG 9 III-C-3, vol. 4229 Folder 19, File 5 Operation Avion 12th CIB S.G. 23/866 "Report on Operations carried out by the 12th Canadian Infantry Brigade between 26th June and 28th June 1917," 3 and 5; LAC RG 9, III-D-3, vol. 4919, WD, 10th Btn., April 1917 Appendix 118, Report on Operations, April 28th, 1917; LAC RG 9, III-D-3, vol. 4871, WD, 2nd Bde., 28 April 1917; LAC RG 41, vol. 11, 25th Btn., Anderson, 5/14-15; LAC RG 41, vol. 12, 31st Btn., W. McCombie-Gilbert, 2/14.

[24] Samuels, *Command or Control?*, 158.

[25] Rawling, *Surviving Trench Warfare*, 88; Samuels, *Command or Control?*, 181.

[26] Stewart, "Attack Doctrine in the Canadian Corps," 139.

[27] Samuels, *Command or Control?*, 186-197; Jack Sheldon, *The German Army at Passchendaele* (Barnsley: Sword & Pen, 2007), xi-xii.

[28] John A. English and Bruce I. Gudmundsson, *On Infantry, Revised Edition* (Westport: Greenwood, 1994), 26.

[29] Lupfer, *Dynamics of Doctrine*, 16; Samuels, *Command or Control?*, 175.

[30] Nicholson, *C.E.F.*, 269; Sheldon, *The German Army on Vimy Ridge*, 338-41.

[31] Paddy Griffith. *Battle Tactics of the Western front*, 6; McEvoy and Finlay, *History of the 72nd Canadian Infantry Battalion*, 74.

[32] LAC RG 9 III-C-3, vol. 4011 Folder 17, File 1, "Operations Somme 1916: Notes on, experience gained, lessons learned," 4.

[33] RG 41, Vol. 15, 49th Battalion, M. Palmer 1/16-17.

[34] MacGowan, Heckbert, and O'Leary, *New Brunswick's 'Fighting 26th'*, 23.

[35] David Charles Gregory Campbell, "The Divisional Experience in the C.E.F.," 309-310; Radley, *We Lead Others Follow*, 114; Rawling, *Surviving Trench Warfare*, 135-6.

[36] Author's emphasis. LAC RG 9 vol. 4199, folder 7, file 11, "First Army, No. 1227(g), 9th May 1917.

[37] For additional instances of the bayonet fighting at Hill 70, see: Beattie, *48th Highlanders of Canada*, 246, 248, and 254; Cook, *Shock Troops*, 276 and 280; Cook, "Politics of Surrender," 644; *Letters From the Front*, 241; Featherstonhaugh,

The 13th Battalion, 199-200; Donald Fraser, *The Journal of Private Fraser,* 305; Singer, *History of the 31st Canadian Infantry Battalion,* 240-242; Y. Lubomyr Luciuk and Ron Soroby, *Konowal: A Canadian Hero* (Kingston: Kashtan Press, 2000; Ron Sorobey "Filip Konowal, VC, The Rebirth of a Canadian Hero, *Canadian Military History* 5, No, 2 (Autumn 1996), 47; LAC RG 9 III-C-3, vol. 4014, Folder 25, File 2 "Operations: Hill 70. (1st Cdn. Div. 1-7-17 to 25-9-17," "Report on the Capture of Hill 70 and Puits 14 Bis by 1st Canadian Division. 15th Aug. 1917;" LAC RG 9 III-C-3, vol. 4014, Folder 26, File 1, "Operations: Hill 70 (1st C. I. Bde) Narratives on," "Report on Operations carried out by 2nd Canadian Infantry Brigade against Hill 70, August 1917.," 13, 15 and 22; LAC RG 9 III-C-3, vol. 4014, Folder 26, File 1, "Operations: Hill 70 (1st C. I. Bde) Narratives on," "7th Battalion report on operations – 15th – 17th August 1917," 3; LAC RG 9, III-D-3, vol. 4917, WD, 7th Btn., August 1917, Appendix D, "Report on Operations - 15th-17th August, 1917.," 3; LAC RG 9, III-D-3, vol. 4930, WD, 20th Btn., 15 August 1917; LAC RG 41, vol. 11, 25th Btn., C. B. Holmes and C. J. Alson, 2/7.

[38] W. D., 31st Btn., September 1917, Appendix B8.

[39] Nicholson, *The Gunners of Canada*, 294.

[40] Ibid, 297.

[41] Featherstonhaugh, *The 13th Battalion*, 197.

[42] LAC RG 9, III-D-3, vol. 4922, WD, 13th Btn., 15 August 1917.

[43] LAC RG 9, III-D-3, vol. 4917, WD, 7th Btn., August 1917, Appendix D, "Report on Operations-15th-17th August, 1917," 3.

[44] LAC RG 9 III, C-3, vol. 4014, Folder 26, File 1, "Operations: Hill 70 (1st C. I. Bde), Narratives on" "Narrative of Operations Against Hill 70, near Loos, in so far as they concern the 4th Canadian Battalion."

[45] Beattie, 248.

[46] RG 9 III-C-3, Vol. 4014, Folder 25, File 2, "Operations: Hill 70. (1st Cdn. Div). 1-7-17 to 25-9-17"

[47] Featherstonhaugh, *The 13th Battalion*, 199-200.

[48] War Office, *Assault Training 1917*, (London: Harrison and Sons, 1917), 8.

[49] LAC RG 9, III-D-3, vol. 4930, WD, 20th Btn., 18 August 1917.

[50] Ibid.

[51] Corrigall, *The History of the Twentieth Canadian Battalion*, 144.

[52] Nicholson, *C.E.F.*, 266; Rawling, *Surviving Trench Warfare*, 141.

[53] Additional instances of Canadian counter-charges are found in: George Stirrett, *A Soldiers Story – 1914-1918*, (Unpublished manuscript, Canadian War Museum)., 11; Alexander McClintock, *Best O' Luck: How a Fighting Kentuckian won the Thanks of Britian's King*, (Ottawa: CEF Books, 2000), 57; H. M. Urquhart, *The History of the 16th Battalion (The Canadian Scottish) Canadian Expeditionary Force in the Great War, 1914-1919*, (Toronto: MacMillan, 1932), 309.

[54] LAC RG 9 III-C-3, vol. 4125, Folder 5, File 1, "Operations: Hill 70 6th Cdn Inf Bde," "Narrative of Operations of 6th Canadian Infantry Brigade," 3.

[55] Singer, *History of the 31st Canadian Infantry Battalion*, 242.

[56] LAC RG 9 III-C-3 Vol 4125 Folder 5, File 1, "Operations: Hill 70 6th Cdn Inf Bde," "Narrative of Operations of 6th Canadian Infantry Brigade," 3.

[57] LAC RG 41, vol. 12, 29th Btn., Robert Hanna V.C., 2/16.

[58] Jack Sheldon, *The German Army at Cambrai* (Barnsley: Pan & Sword, 2009), 1-2.

[59] Original emphasis. LAC RG 9 III-C-1, vol. 3859 Folder 85, File 5, "Lessons to be drawn from recent offensive operations: 18-5-17 to 29-9-17," "First Army No. 1227 (s)"

[60] Corrigall, *The History of the Twentieth Canadian Battalion*, 162-3; McEvoy and Finlay, *History of the 72nd Canadian Infantry Battalion*, 79; Russenholt, *Six Thousand Canadian Men*, 124; Shackleton, *Second to None*, 177; Singer, *History of the 31st Canadian Infantry Battalion*, 268-9; Letter from Lt. Don Cameron to Miss Gissing, University of Toronto archives, A73-0026/358 (12); LAC RG 9 III-C-3, vol. 4132 Folder 12, File 8, (Headquarters, 6th Brigade), "Narrative Report of Operations for Capture of Passchendaele," 5, 6, and 7; LAC RG 41, vol. 13, 43rd Btn., D. Mowat, 1/9-10; LAC RG 41, vol. 14, 49th Btn., A Black and B. Morrisson 2/3-4.

[61] McWilliams and Steel, *The Suicide Battalion*, 116.

[62] Ibid.

[63] Griffith, *Battle Tactics*, 125.

[64] LAC RG 9, III-D-3, vol. 4947, WD, 4 CMR, Appendix 1, "Narrative of Events of the action of October 26th 1917," 2.

[65] Jason Adair, "The Battle of Passchendaele, The experiences of Lieutenant Tom Rutherford, 4th Battalion, Canadian Mounted Rifles," *Canadian Military History* 13, No. 4, Autumn 2004. 70-71.

[66] Adair, "The Battle of Passchendaele," 70-71.

[67] LAC RG 9, III-D-3, vol. 4944, WD, 85th Btn., October 1917, Appendix A, "Report on move of Battalion into the line and report on operation," 3.

[68] Joseph Hayes, *The Eighty-Fifth in France and Flanders*, 97.

[69] Ibid., 101.

[70] LAC RG 9, III-D-3, vol. 4944, WD, 85th Btn., October 1917, Appendix A, "Report on move of Battalion into the line and report on operation," 3.

[71] Joseph Hayes, *The Eighty-Fifth in France and Flanders*, 91.

[72] Ibid, 97.

[73] WD, 85th Btn., October 1917, Appendix A, "Report on move of Battalion into the line and report on operation," 3.

[74] W.D., 85th Canadian Infantry Battalion, October 1917, Appendix A, "Report on move of Battalion into the line and report on operation."

[75] Hayes, 100-1.

[76] For additional accounts of close combat in Canadian operations around Passchendaele see: Bird, *Ghosts Have Warm Hands*, 91-2; D. J. Goodspeed, *Battle Royal*, 205-206; Singer, *History of the 31st Canadian Infantry Battalion*, 269; LAC RG 9, III-C-3, vol 4132 Folder 12, File 8 Headquarters, 6th Canadian Infantry Brigade, Narrative Report of Operations for Capture of Passchendaele," Nov 6th, 1917, 5.

[77] The edition of *Bayonet Fighting Illustrated 1917* at the Canadian War Museum includes a handwritten dedication dated August 1917. This pamphlet was also reprinted for use by the American Army in August 1917: Army War College, *Notes on Bayonet Training, No. 2, adapted from a Canadian publication*, (Washington: Government Printing Bureau, 1917).

[78] *Bayonet Fighting Illustrated 1917*, 15-20.

[79] Ibid, 94

[80] Ibid, 57.

[81] Ibid, 58-9.

[82] *Assault Training 1917*, 3.

[83] *Assault Training 1917*, 11-12. This same exercise is found published separately as early as 25 March 1917, see: LAC RG 9 III, vol. 3870, Folder 112 File 11, memo dated 25 March 1917.

[84] LAC RG 9 III, vol. 3870, Folder 112, File 11, General Staff memo dated March 25th, 1917. and LAC RG 9 III, 3870 Folder 112, File 13, War Office, *S.S. 152 System of Testing & Training Reinforcements: Carried out by Base Training Schools, B.E.F.* (Army Printing Services, February 1917)

[85] Rawling, *Surviving Trench Warfare*, 131.

Chapter VII: 1918, The Bayonet and the War of the Movement

[1] Cook, *Shock Troops*, 372

[2] Ibid., 373

[3] Ibid., 393; Nicholson, *C. E. F.*, 369-72 and 378-85..

[4] Campbell, "Divisional Experience in the C. E. F....," 369-70; Cook, *Shock Troops*, 395-6

[5] Instances of bayonet use during this period are found in: Bernard McEvoy and A. H. Finlay, History of the 72nd Canadian Infantry Battalion: Seaforth Highlanders of Canada, (Vancouver: Cowan and Brookhouse, 1920), 97; D. G. Scott Calder, The History of the 28th (Northwest) Battalion, C.E.F. (October 1914-June 1919), (Unpublished History, Fort Frontenac Library), 187; H. R. N. Clyne, Vancouver's 29th: A Chronicle of the 29th in Flanders Fields, (Vancouver: Tobin's Tigers Association, 1964), 133; R. C. Featherstonhaugh, *The 13th Battalion Royal Highlanders of Canada: 1914-1919,* (The 13th Battalion, Royal Highlanders of Canada, 1925), 236; R. C. Featherstonhaugh, *The 24th Battalion, C.E.F., Victoria Rifles of Canada 1914-1919*, (Montreal: Gazette Printing, 1930), 208; H. C. Singer, History of the 31st Canadian Infantry Battalion C.E.F., 321-2, 332, and 333; RG 9 III-C-3, vol 4106 Folder 22, file 14 Operations Minor (6th CIB 6th Canadian Infantry Brigade, Minor Operation by 31st (Alberta) Bn. Neuville Vitasse night 21st/22nd May 1918. 4 and 5; RG 9 III-C-3, vol 4106 Folder 22, file 14 Operations

Minor (6th CIB "Narrative of Minor operation Carried out on night June 2nd/3rd opposite front of mercatel sector by 6th Canadian Infantry Brigade--29th (Vancouver) Bn. June 2/3. 1918; RG 41, Volume 15 50th Battalion Mr A. Turner Tape 4, page 1

[6] McEvoy and Finlay, *History of the 72nd Canadian Infantry Battalion*, 90.

[7] LAC, RG 41, vol. 17, 116th Battalion, General Pearkes, 4/8-9

[8] RG 9, III 3108, T-4-37, Oct 1918, 31½ hours of bayonet training in a 10 week course. As opposed to 79½ Musket and 10½ bombing.

[9] Cook, *Shock Troops*, 410;

[10] Nicholson, *C.E.F.*, 393-6.

[11] Instances of Bayonet Fighting during the Amiens operations not included in this chapter see: Bird, *Ghosts Have Warm Hands,* 157; Dinesen, *Merry Hell!,* 239; Featherstonhaugh, *The 13th Battalion*, 252; ---, *The 14th Battalion,* 219; ---, *The 24th Battalion,* 226; Goodspeed, *Battle Royal,* 237 and 239; Johnson, *The 2nd Canadian Mounted Rifles,* 67; J. F. B. Livesay, *Canada's Hundred Days: With the Canadian Corps from Amiens to Mons, Aug. 8-Nov. 11, 1918* (Toronto: Thomas Allen, 1919), 43, 47, 65, 66, and 78; McEvoy and Finlay, *History of the 72nd Canadian Infantry Battalion,* 120 and 121; Nicholson, *C.E.F.*, 375 and 392; Shackleton, *Second to None,* 244; Singer, *History of the 31st Canadian Infantry Battalion,* 344, 352, 362 and 379-81; Tascona and Wells, *Little Black Devils,* 111-2 and 112; Topp, *The 42nd Battalion,* 224 and 233; *Canada in the Great World War, Volume VI*, 274, 281 and 298; LAC RG 9 III-C-3, vol. 4089, Folder 5, file 1, "operations: Amiens (1st Division Report) 8-8-18 to 20-8-18" 16th Cdn Battalion, 3rd CIB "1st Canadian Division Report on Amiens Operations," 4; LAC RG 9 III-C-3, vol. 4015, Folder 30, file 6, "Operations Amiens 2nd Bde," "2nd Canadian Infantry Brigade narrative of the operations east of Amiens August 8/9 1918," 20; LAC RG 9, III-D-3, vol. 4938, WD, 42nd Btn., 12 August 1918 and Appendix 2 "Report on operations carried out by the 42nd Canadian Battalion, R. H. C. on the PARVILLIERS SECTOR on the 12th, 13th, 14th, and 15th of August 1918," 1 and 29; LAC RG 41, vol. 11, 27th Btn., John Nind,10; LAC RG 41, vol. 12, 28th Btn., Arthur B. Goodmurphy, 2/17; LAC RG 41, vol. 12, 31st Btn., D. J. Avison, 2; LAC RG 41, vol. 14, 46th Btn., Bob Bron, 3/1; LAC RG 41, vol. 14, 46th Btn., Crowe, 2/10; LAC RG 41, vol. 15, 50th Btn., Mr. A. Turner, 3/3 and 6; LAC RG 41, vol. 11, 27th Btn., Mr. Fryday, 2/19.

[12] The Adjutant, *The 116th Battalion in France* (Toronto: E. P. S. Allen, 1921), 68.

[13] LAC RG 41, vol. 16, 78th Btn., Major F. G. Thompson, 2/11.

[14] LAC RG 9, III-D-3, vol. 4900, WD, 9th Bde., Appendix 11, "Narrative of Operation, from August 8th to 16th, 1918," 3.

[15] LAC RG 9, III-D-3, vol. 4938, WD, 42nd Btn., August 1918, Appendix 1, "Report of operations carried out by the 42nd Canadian Battalion R.H.C. on the 8th August 1918 at Hill 102."

[16] Topp, *The 42nd Battalion*, 211-212.

[17] LAC RG 9, III-D-3, vol. 4938, WD, 42nd Btn., August 1918, Appendix 1, "Report of operations carried out by the 42nd Canadian Battalion R.H.C. on the 8th August 1918 at Hill 102."

[18] Topp, *The 42nd Battalion*, 210.

[19] LAC RG 9, III-D-3, vol. 4938, WD, 42nd Btn., Appendix 1, "Report of operations carried out by the 42nd Canadian Battalion R.H.C. on the 8th August 1918 at Hill 102," 1-2.

[20] Topp, *The 42nd Battalion*, 212-3.

[21] Other reference to bayonet charges on Artillery batteries in the Hundred Days see: Featherstonhaugh, *The 13th Battalion*, 253; McEvoy and Finlay, *History of the 72nd Canadian Infantry Battalion*, 142; Urquhart, *The History of the 16th Battalion*, 277; Joseph Hayes, *The Eighty-Fifth in France and Flanders*, 132; LAC RG 41, vol. 15, 50th Btn., Mr A. Turner, 3/6; LAC, RG41, vol. 11, 27th Battalion A D McEtherenm,1/22,

[22] LAC RG 9, III-D-3, vol. 4945, WD, 102nd Btn., August 1918, Appendix B, "Report on operations," 1.

[23] Ibid, 1.

[24] Rawling, *Surviving Trench Warfare*, 182.

[25] LAC RG 9 III-C-3, vol. 3859, folder 85, file 7, "Lessons learned from Recent Fighting Referred to by the Army Commander at a Conference held at Canadian Corps HQ, 30 August, 1918."

[26] LAC RG 9, III, vol. 3859, folder 85, file 7, "Lessons learned from Recent Fighting Referred to by Army Commander at a Conference held at Canadian Corps HQ," cited in Dean Chappelle, "The Canadian Attack at Amiens, 8-11 August 1918," *Canadian Military History*, vol. 2, no. 2 (1993), 93. Also see: Rawling, *Surviving Trench Warfare*, 83 and 183.

[27] Rawling, *Surviving Trench Warfare*, 207.

[28] LAC RG 9, III-D-3, vol. 4944, WD, 78th Btn., Appendix 4, "Report of operations carried out by the 78th Canadian Infantry Battalion (Winnipeg Grenadiers) from 8th to 12th August 1918 inclusive," 1

[29] Livesay, *Canada's Hundred Days*, 83-4; LAC RG 41, vol. 16, 78th Btn., Oscar Ericson, 2/4-5.

[30] Featherstonhaugh, *The 13th Battalion*, 252-3; Goodspeed, *Battle Royal*, 239; Joseph Hayes, *The Eighty-Fifth in France and Flanders*, 133; Shackleton, *Second to None*, 244; Tascona and Wells, *Little Black Devils*, 111-2; *Canada in the Great World War, Volume VI*, 274 and 298; LAC RG 9 III-C-3, vol. 4015 Folder 30, file 6 Operations Amiens 2nd Bde "2nd Canadian Infantry Brigade narrative of the operations east of Amiens August 8/9 1918, 8-9 and 20; LAC RG 9 III-C-3 vol. 4089 Folder 5, file 1 "operations: Amiens (1st Division Report) 8-8-18 to 20-8-18 (16th Cdn Battalion, 3rd CIB)," "1st Canadian Division Report on Amiens Operations," 4; LAC RG 9, III-D-3, vol. 4915, WD, 3rd Btn., August 1918, Appendix 1 "Narrative of Operations, August 6th to 9th, 1918," 3; LAC RG 9, III-D-3, vol. 4938, WD, 42nd Btn., August 1918, Appendix 2, "Report on operations carried out by the 42nd Canadian Battalion, R. H. C. on the PARVILLIERS SECTOR on the 12th, 13th, 14th, and 15th of August 1918," 1; LAC RG 41, vol. 11, 27th Btn., John Nind, 2/1; LAC RG 41, vol. 15, 50th Btn., Mr. A. Turner, 3/3.

[31] Johnson, 68; Other records of bayonet fighting occurring at Orange Hill and Chappel Hill see: RG 9 III-C-3, vol. 4016 Folder 31, file 1 "Operations Arras (1st Cdn, Div) Narrative" Arras Operations Night 28th/29th August to night 4th/5th September 1918.

[32] WD, 7th Bde., 26 August 1918.

[33] Featherstonhaugh, *The Royal Canadian Regiment*, 342-3.

[34] WD, 7th Bde., 26 August 1918.

[35] Featherstonhaugh, *The Royal Canadian Regiment*, 342-3.

[36] LAC RG 9, III-D-3, vol. 4912, WD, PPCLI, August 1918, Appendix 2, "Narrative of Operations - August 25th/29th 1918," 1.

[37] LAC RG 9, III-D-3, vol. 4894, WD, 7th Bde., 26 August 1918.

[38] Lynch, *Princess Patricia's Canadian Light Infantry*, 126.

[39] Ibid, 126.

[40] Ibid, 127.

[41] *Assault Training 1917*, 12.

[42] Lynch, *Princess Patricia's Canadian Light Infantry,* 126-8.

[43] Ibid, 126-8.

[44] Ibid, 126-8.

[45] *Bayonet Training 1916*, 11, para. 32. An Officer, *Practical Bayonet Fighting,* 14, para. 19.

[46] Lynch, *Princess Patricia's Canadian Light Infantry,* 127-8.

[47] *Bayonet Fighting Illustrated 1917*, 24; *Bayonet Training 1916*, para. 23.

[48] Instances of bayonet fighting in the fighting between August 26th and September 2nd 1918 not investigated in this chapter, see: anon. *Extracts from Princess Patricia's Canadian Light Infantry 1914-1919* (Ottawa: Queen's Printer, 1959), 73; Bird, *Ghosts Have Warm Hands,* 173-4 and 174-5; Featherstonhaugh, *The 24th Battalion*, 237; ——, *The Royal Canadian Regiment*, 343; ——, *The 13th Battalion*, 267; Joseph Hayes, *The Eighty-Fifth in France and Flanders,* 139, 140, 143 and 149; Livesay, *Canada's Hundred Days,* 162, 163; McEvoy and Finlay, *History of the 72nd Canadian Infantry Battalion,* 128; Nicholson, *C.E.F.*, 408 and 411; Singer, *History of the 31st Canadian Infantry Battalion,* 379-81; LAC RG 9 III-C-3, vol. 4016 Folder 31, file 1 "Operations Arras (1st Cdn, Div) Narrative" Arras Operations Night 28th/29th August to night 4th/5th September 1918;" LAC RG 9 III-C-3, vol. 4016, Folder 31, File 2, "2nd Canadian Infantry Brigade Narrative of Operations East of Arras August 25th to September 3rd, 1918," 14, 20 and 22; LAC RG 9 III-C-3, vol. 4089, Folder 5 file 3, "Operations: Arras – Drocourt Queant Line 28-8-18 to 4-9-18," "Arras Operations. Night 28th.29th August to 4th/5th September 1918," Section 2, 2; LAC RG 9, III-D-3, vol. 4944, WD, 85th Btn., September 1918, Appendix "Scarpe," "Scarpe Operation 30th Aug. - 3rd Sept., 1918," 3 and 4; LAC RG 41, vol. 11, 27th Btn., John Nind, 10.

[49] LAC RG 9 III-C-3, vol. 4016 Folder 31, File 7, "Operations Scarpe and Drocourt-Queant Line (3rd Cdn. Inf. Brigade) 26-8-18 to 2-9-18," 3.

[50] McKean, *Scouting Thrills,* 211-2

[51] LAC RG 9 III-C-3, vol. 4016 Folder 31, File 7, "Operations Scarpe and Drocourt-Queant Line (3rd Cdn. Inf. Brigade) 26-8-18 to 2-9-18, 2.

[52] Ibid.; LAC RG 9, III-D-3, vol. 4923, WD, 14th Btn., September 1918, Appendix 5. "Report on Operations September 1st," 3.

[53] Ibid, 2.

[54] LAC RG 9, III-D-3, vol. 4922, WD 13th Btn., 2 September 1918.

[55] Ibid.

[56] LAC RG 9, III-D-3, vol. 4923, WD, 14th Btn., Appendix 5, "Royal Montreal Regiment, Report on Operation of September 2nd, 1918," 1.

[57] LAC RG 9, III-D-3, vol. 4923, WD, 14th Btn., 2 September 1918.

[58] LAC RG 9, III-D-3, vol. 4923, WD, 14th Btn., Appendix 5, "Royal Montreal Regiment, Report on Operation of September 2nd, 1918," 1

[59] Ibid, 1

[60] LAC RG 9, III-D-3, vol. 4822, WD, 3rd Bde., September 1918, Appendix 3a, "Narrative of Events 26th August to 3rd September," Appendix 3, "13th Battalion report on DROCOURT-QUEANT Operation."

[61] Ministry Overseas Military Forces of Canada, *Report of the Overseas Military Forces of Canada, 1918*, (London, 1918), 141, 168 and 184.

[62] For accounts of bayonet fighting occurring between September 2nd and November 11th 1918 see: Bird, *Ghosts Have Warm Hands*, 203; Cook, *Shock Troops*, 538; Featherstonhaugh, *The 13th Battalion*, 288; W. L. Gibson, *Records of the Fourth Canadian Infantry Battalion In the Great War 1914-1918*. (Toronto: Maclean Publishing, 1924), 234, 235, 247 and 248; Goodspeed, *Battle Royal*, 256; Joseph Hayes, *The Eighty-Fifth in France and Flanders*, 171, 179, and 207; Johnson, *The 2nd Canadian Mounted Rifles*, 70; Livesay, *Canada's Hundred Days*, 239, 269, 322, 360; MacGowan, Heckbert, and O'Leary, *New Brunswick's 'Fighting 26th'*, 301; McEvoy and Finlay, *History of the 72nd Canadian Infantry Battalion*, 144; McWilliams and Steel, *The Suicide Battalion*, 197; Topp, *The 42nd Battalion*, 259; *Canada in the Great World War, Volume VI*, 294; LAC RG 9 III-C-3, vol. 4028, Folder 16, file 2, "Operations: Canal du Nord: 1st Bde" "Narrative of Operations, September 27th, 1918, The assault of the Canal DU NORD and subsequent attack on DELIGNY MILLRIDGE (Northwest of the Village of BOURLON) by the 4TH CANADIAN INFANTRY BATTALION," 2 and 3; LAC RG 9 III-C-3, vol. 4028 Folder 16, file 3, "Operations: Cambrai, 1st Bde." ref. 671-8 October 10th, 1918, 1st Battalion Narrative Oct. 1st, 1918; LAC RG 9 III-C-3, vol. 4028 Folder 16, file 4 "Operations: Attack on Abancourt" "Narrative of Operations, October 1st, 1918, The attack on Abancourt by 4th Canadian Infantry Battalion;" LAC RG 9 III-C-3 vol. 4089 Folder 5, file 4 "1st Canadian Division:

Report on Canal du Nord -Bourlon Wood – Cambrai Operations" Section 3, page 7; LAC RG 9 III-C-3, vol. 4230 folder 22, file 4 Operations: Valenciennes, 12th Inf Bde 85th Canadian Infantry Battalion, Nova Scotia Highlanders, Report on Operations from Oct 22nd until Nov. 6th, 1918, in the VALENCIENNES Operation 6-11-18;" LAC RG 9, III-D-3, vol. 4917, WD, 4th Btn., 1 October 1918; LAC RG 41, vol. 14, 46th Btn., Lt. Col J. L. Hart, 3/3; LAC RG 41, vol. 18, Fort Gary Horse, Lt. Col. Strachan 1/9.

Conclusion

[1] LAC RG 9, III-D-3, vol. 4938, WD, 42nd Btn., Appendix 1, "Report of operations carried out by the 42nd Canadian Battalion R.H.C. on the 8th August 1918 at Hill 102."

[2] LAC RG 9, III-D-3, vol. 4944, WD, 85th Btn., October 1917, Appendix A, "Report on move of Battalion into the line and report on operation," 3.

[3] LAC RG 9, III-D-3, vol. 4916, WD, 5th Btn., May 1915, Appendix A, "Festubert."

[4] LAC RG 9, III-D-3, vol. 4944, WD, 85th Btn., September 1918, Appendix "Scarpe," "Scarpe Operation 30th Aug. - 3rd Sept, 1918," 4.

[5] LAC RG 9, III-D-3, vol. 4944, WD, 78th Btn., Appendix 4, "Report of operations carried out by the 78th Canadian Infantry Battalion (Winnipeg Grenadiers) from 8th to 12th August 1918 inclusive,"1

[6] LAC RG 9, III-D-3, vol. 4947, WD, 2 CMR, Sept 1916, "Report on Operations From Sept 27th to Oct 2nd, 1916." LAC RG 9, III-D-3, vol. 4944, WD, 85th Btn., October 1917, Appendix A, "Report on move of Battalion into the line and report on operation," 3. LAC RG 9, III-D-3, vol. 4944, WD, 85th Btn., September 1918, Appendix "Scarpe," "Scarpe Operation 30th Aug.-3rd Sept, 1918," 4.

[7] LAC RG 9, III-D-3, vol. 4944, WD, 87th Btn., October 21st, 1916. LAC RG 9, III-D-3, vol. 4944, WD, 85th Btn., September 1918, Appendix "Scarpe," "Scarpe Operation 30th Aug.-3rd Sept, 1918," 4.

[8] LAC RG 9, III-D-3, vol. 4937, WD, 31st Btn., April 1917, Appendix C-2, "Report on Operations," 1.

[9] LAC RG 9, III-D-3, vol. 4922, WD, 13th Btn., 15 August 1917.

[10] LAC RG 9, III-D-3, vol. 4911, WD, RCR, April 1917, Appendix 2, "Summary of Operations," 2.

[11] LAC RG 9, III-D-3, vol. 4938, WD, 42nd Btn., 16 November 1917.

[12] LAC RG 9, III-D-3, vol. 4924, WD, 15th Btn., August 1917, Appendix 1, "Note on the Operations against HILL 70 on August 15th & 16th, 1917," 3.

[13] Original emphasis. D.H.H., 163.009 (D99), G. Dalby, "Training in the use of the bayonet," *Cavalry Journal*, vol. XI, No. 41 (July 1921), 2.

[14] Ibid., 2-3.

[15] LAC RG 41, vol. 10, 21st Btn., W.P. Doolan, 2/1.

Works Cited

Archival Sources

Library and Archives Canada:

RG 9, Department of Militia and Defence Records (CEF war diaries from RG 9 are posted on line at www.collectionscanada.ca/02/020152_e.html)

RG 24, Department of National Defence Records.

RG 41, Canadian Broadcasting Corporation Records

Canadian Letters & Images Project <http://www.canadianletters.ca/>

Directorate of History and Heritage:

163.009 (D99). Dalby, G. "Training in the use of the bayonet." *Cavalry Journal.* Vol. XI, No. 41 (July 1921). 2. D.H.H..

Manuals and Pamphlets

Adjutant Genera"'s Office. *Bayonet exercise.* London: H.M.S.O., 1860.

An Officer. *Practical Bayonet Fighting: With Service Rifle and Bayonet.* London: The Bazaar, Exchange and Mart Office, Windsor House, Bream's Buildings, E.C.. 1915.

Angelo, Charles Henry. *Angelo's Bayonet Exercise.* London: Parker, Furnivall, and Parker, Military Library, 1853.

Burton, Richard F. *A Complete System of Bayonet Exercise.* London: William Clowes and Sons, 1853).

Department of Militia and Defence. *Infantry Training 1904.* Ottawa: Government Printing Bureau, 1904.
 ---. *The Organization of Bayonet Fighting and Physical Training in a Battalion C.E.F. (Revised) 1916.* Ottawa: Government Printing Bureau, 1916.

Headquarters Canadian Overseas Military Forces. *Bayonet Fighting Illustrated 1917.* London: Harrison and Sons, 1917.

McClaglen, Leopold. *Bayonet Fighting for War.* London: Harrison and Sons, 1916.

War Office. *Angelo's bayonet exercise.* London: John W. Parker and Son, 1849.
 ---. *Angelo's bayonet exercise.* London: John W. Parker and Son, 1857.
 ---. *Infantry sword and carbine sword-bayonet exercises.* London: H.M.S.O., 1880.
 ---. *Infantry Training 1902.* London: Eyre & Spottiswood, 1902.
 ---. *Infantry Training 1905.*
 ---. *Instruction in Bayonet Fighting.* London: Harrison and Sons, 1907.
 ---. *Infantry Training 1905.* London: Wyman and Sons, 1908.
 ---. *Infantry Training 1911.* London: Wyman and Sons, 1911.
 ---. *Infantry Training 1914.* London: Harrison & Sons, 1915.
 ---. *Bayonet Fighting: Instruction with Service Rifle and Bayonet: 1915.* London: Harrison and Sons, 1915.
 ---. *Bayonet Training 1916 (Provisional).* London: Harrison and Sons, 1916.

—. *Bayonet Training 1916.* London: Harrison and Sons, 1916.

—. *Army Council Instruction 1916.* London: Harrison and Sons, 1916.

—. *Assault Training 1917.* London: Harrison and Sons, 1917.

—. *System of Testing & Training Reinforcements: Carried out by Base Training Schools, B.E.F.* Army Printing Services, February 1917

—. Methods of Unarmed Attack and Defence. London: Harrison and Sons, 1917.

Monographs

Anon. *Extracts from Princess Patricia's Canadian Light Infantry 1914-1919.* Ottawa: Queen's Printer, 1959.

Anon. *Letters From the Front: Being a Record of the art played by Officers of the Bank in the Great War 1914-1919; Volume 1.* Edited by Charles Lyons Foster. Toronto: Southam Press, 1920.

Amberger, J. Christoph. *The Secret History of the Sword: Adventures in Ancient Martial Arts.* Burbank: Hammerterz Forum, 1996.

Angelo, Henry. *The Reminiscences of Henry Angelo.* Manchester: Ayer Publishing, 1972.

Anglo, Sydney. *The Martial Arts of Renaissance Europe.* New Haven: Yale University Press. 2000.

Bailey. Johnathan. "British Artillery in the Great War." *British Fighting Methods in the Great War.* ed. Paddy Griffith. Portland: Frank Cass. 1996.

Baldwin. Harold. *Holding the Line.* London: George J. McLeod. 1918.

Barnett, Corelli. *The Sword Bearers, Studies in Supreme Command in the First World War.* Toronto: Hodder & Stoughton, 1986.

Beattie. Kim. *48th Highlanders of Canada. 1891-1928.* Toronto: Southam Press, 1932.

Beckett, Ian F. W. *The Great War 1914-1918.* New York: Pearson Education, 2001.

Beckett, Ian F. W. and Simpson, Keith. *A Nation in Arms: A Social Study of the British Army in the First World War.* Dover: Manchester University Press, 1985.

Berton, Pierre. *Vimy.* Toronto: Anchor, 2001.

Bidwell, Shelford and Graham, Dominick. *Firepower, British Weapons and Theories of 1904-1945.* Boston: George Allen & Unwin, 1982.

Bird, Will R. *Ghosts Have Warm Hands.* Vancouver: Clarke, Irwin & company Ltd., 1968.

Bond, Brian. *War and Society in Europe, 1870-1970.* Montreal: McGill-Queen's University Press, 1998.
---. *The Unquiet Western Front, Britain's role in literature and history.* New York: Cambridge University Press, 2002.

Boot, Max. *War Made New: Technology, Warfare, and the Course of History, 1500 to Today.* New York: Gotham, 2006.

Bourke, Joanna. *An Intimate History of Killing, Face-to-Face Killing in Twentieth Century Warfare.* London: Granta, 1999.

Bowman, Timothy. *The Irish Regiments in the Great War: Discipline and Morale.* New York: Manchester University Press, 2006.

Bruce, Robert B. *Petain: Verdun to Vichy.* Dulles: Brassey's, 2008.

Bryson, Richard "The Once and Future Army." *Look to your front, Studies in the First World War.* Brian Bond et al. Staplehurst: Spellmount, 1999.

Burg, David F. and Purcell, L. Edward. *Almanac of World War I.* Lexington: University Press of Kentucky, 2004.

Calder, D.G. Scott. *The History of the 28th (Northwest) Battalion, C.E.F. (October 1914-June 1919).* Regina: Regina Rifle Regiment, 1961.

Cassar, George H. *The Tragedy of Sir John French.* Mississauga: University of Delaware Press, 1985.

Castle, Edgerton. *Schools and Masters of the Fence: From the Middle Ages to the Eighteenth Century.* London: Arms and Armour Press, 1969.

Clausewitz, Carl von. *On War*. Tr. Peter Paret. Princeton: Princeton University Press, 1976.

Clifford, W. G. *The British Army*. Alcester: Read Books, 2008.

Clyne, H. R. N. *Vancouver's 29th: A Chronicle of the 29th in Flanders Fields*. Vancouver: Tobin's Tigers Association, 1964.

Connelly, Mark. *Steady the Buffs!: Regiment, a Region, and the Great War*. New York: Oxford University Press, 2006.

Cook, Tim. *No Place to Run, The Canadian Corps and Gas Warfare in the First World War*. Toronto: UBC Press, 1999.
———. *Clio's Warriors, Canadian Historians and the writing of the World Wars*. Vancouver: UBC Press, 2006.
———. *At the Sharp End, Canadians Fighting the Great War 1914-1916, Volume One*. Toronto: Viking Canada, 2007.
———. *Shock Troops, Canadians Fighting the Great War 1917-1918, Volume 2*. Toronto: Viking Canada, 2008.

Corrigall, D. J. *The History of the Twentieth Canadian Battalion (Central Ontario Regiment) Canadian Expeditionary Force in the Great War, 1914-1918*. Toronto: Stone & Cox, 1935.

Corvisier, André; Childs, John; and Turner, Chris. *A Dictionary of Military History and the Art of War*. Cambridge: Blackwell Publishing, 1994.

Dinesen, Thomas. *Merry Hell! A Dane with the Canadians*. London: Jarrolds, 1930.

Doughty, Robert A. *Pyrrhic Victory: French Strategy and Operations in the Great War*. Cambridge: Harvard University Press, 2005.

Duffy, Christopher. *Through German Eyes, The British and the Somme 1916*. London: Weidenfeld & Nicolson, 2006.

Duguid, A. Fortescue. *The Official History of Canadian Forces in the Great War: Volume I*, Ottawa: J. O. Patenaude, 1938.
———. *History of the Canadian Grenadier Guards: 1760-1964*, Montreal: Gazette Printing, 1965.

Echevarria II, Antulio J. *After Clausewitz, German Military Thinkers Before the Great War*. Lawrence: University Press of Kansas, 2000.

Echevarria, Antulio Joseph. *Imagining Future War: The West's Technological Revolution and Visions of Wars to Come, 1880-1914*. Westport: Greenwood Publishing Group, 2007.

Ellis, John. *Eye-Deep in Hell*. London: Croom Helm, 1976.
---. *The Social History of the Machine Gun*. Baltimore: Johns Hopkins University Press, 1986.

English, John A. and Gudmundsson, Bruce I. *On Infantry, Revised Edition*. Westport: Greenwood, 1994.

Featherstonhaugh, R. C... *The 13th Battalion Royal Highlanders of Canada: 1914-1919*. Canada: The 13th Battalion, Royal Highlanders of Canada, 1925.
---. *The Royal Montreal Regiment 14th Battalion, C. E. F. 1914-1925*. Montreal: Gazette Publishing, 1927.
---. *The 24th Battalion, C.E.F., Victoria Rifles of Canada 1914-1919*. Montreal: Gazette Printing, 1930.
---. *The Royal Canadian Regiment: 1883-1933*. Montreal: Gazette Publishing, 1936.

Fraser, Donald. *The Journal of Private Fraser: 1914-1918 Canadian Expeditionary Force*. Edited by Reginald H. Roy. Victoria: Sono Nis Press, 1985.

Fuller, William C. Jr. "What is a Military Lesson?" *Strategic Logic and Political Rationality: Essays in Honor of Michael I. Handel*. Ed. Bradford A. Lee and Karl-Friedrich Walling. London: Routledge, 2003.

Gibson, W. L. *Records of the Fourth Canadian Infantry Battalion In the Great War 1914-1918*. Toronto: Maclean Publishing, 1924.

Gilchrist, Harry L. *A Comparative Study of World War Casualties: from Gas and Other Weapons*. Washington: U. S. Government Printing Office, 1928.

Goodspeed, D. J. *Battle Royal: A History of the Royal Regiment of Canada 1862-1962*. Toronto: Charters Publishing, 1962.

---. *The Road Past Vimy: The Canadian Corps 1914-1918*. Toronto: MacMillan of Canada, 1969.

Gole, Henry G. and Stofft, William A. *General William E. DePuy: Preparing the Army for Modern War*. Lexington: University Press of Kentucky, 2008).

Graham, Dominick. "Observations of the Dialectics of British Tactics, 1904-45," from *Men Machines & War*. Ed. Ronald Haycock and Keith Neilson, Waterloo: Wilfrid University Press, 1988.

Gray, John G. *Prophet in Plimsoles: An Account of the Life of Colonel Ronald B. Campbell*. Edinburgh: Edina Press, 1976.

Griffith, Paddy. *Battle Tactics of the Western Front, The British Army's Art of Attack, 1916-1918*. New Haven: Yale University Press, 1994.
    ---. "The Extent of Tactical Reform in the British Army." *British Fighting Methods in the Great War*. Portland: Frank Cass, 1996.

Grossman, David. *On Killing: The Psychological Cost of Learning to Kill in War and Society*. Boston: Back Bay Books, 1996.

Gudmundsson, Bruce I. *Stormtroop Tactics, Innovation in the German Army, 1914-1918*. New York: Praeger, 1989.

Habeck, Mary R. "Technology in the First World War: The View from Below." *The Great War and the Twentieth Century*. Ed. Winter, Jay; Parker, Geoffrey; and Habeck, Mary R. New Haven: Yale University Press, 2000.

Hamilton, Richard F. and Herwig, Holger H. *The Origins of World War I*. New York: Cambridge University Press, 2003.

Harris, John. *The Somme: Death of a Generation*. London: White Lion, 1966.

Harris, Stephen J. *Canadian Brass: The Making of a Professional Army, 1860-1939*. Toronto: University of Toronto, 1988.

Harrison, George Yale. *Generals Die in Bed*. Toronto: Annick Press, 2004.

Haycock, Ronald G. *Sam Hughes: The Public Career of a Controversial Canadian 1885-1916*. Waterloo: Wilfrid Laurier Press, 1986.

Hayes, Joseph. *The Eighty-Fifth in France and Flanders*. Halifax: Royal Print & Litho, 1920.

Herrmann, David G. *The Arming of Europe and the Making of the First World War*. Princeton: Princeton University Press, 1996.

Hewitt, G. E. *The Story of the 28th Battalion, 1914-1917*. London: Canadian War Records Office, 1917.

Holland, J. A. *The Story of the 10th Battalion, 1914-1917*. London: Canadian War Records Office, 1917.

Holley, I. B. Jr. *Technology and Military Doctrine, Essays on a Challenging Relationship*. Maxwell: Air University Press, 2004.

Howard, Michael. "Europe on the Eve of the First World War." *The Coming of the First World War*. Ed. Evans, Robert John Weston and Strandmann, Hartmut Pogge von. New York: Oxford University Press, 1991.

House, Jonathan M. "The Decisive Attack: A New Look at French Infantry Tactics on the Eve of World War I." *Military Affairs*. Vol. 40, No. 4 (Dec., 1976).

Humphries, Marc Osborne. "'Old Wine in New Bottles,' A Comparison of British and Canadian Preparations for the Battle of Arras." *Vimy Ridge, A Canadian Reassessment*. Ed. Hayes, Geoffery; Iarocci, Andrew; and Bechthold, Mike. Waterloo: Wilfrid Laurier University Press, 2007.

Hutton, Alfred. *Bayonet Fencing and Sword Practice*. London: W. Clowes and Sons, 1882.
    ----. *Fixed Bayonets*. London: W. Clowes, 1890.

Iarocci, Andrew. *Shoestring Soldiers: The 1st Canadian Division at War, 1914-1915*. Toronto: University of Toronto Press, 2008.

Jarymowycz, Roman Johann and Starry, Donn A. *Cavalry from Hoof to Track*. Westport: Greenwood Publishing Group, 2008.

Johnson, G. Chalmers. *The 2nd Canadian Mounted Rifles [British Columbia Horse] In France and Flanders*. Vernon: CEF Books, 2003.

Jones, Archer. *The Art of War in the Western World*. University of Illinois Press, 2001.

Keegan, John. *The Face of Battle: A Study of Agincourt, Waterloo and the Somme*. New York: A. Knopf, 1976.
---. *The First World War*. New York: Viking, 1998.

Keene, Louis. *Crumps: The Plain Story of a Canadian who went*. Boston: Houghton Mifflin, 1917.

Kramer, Alan. *The Dynamic of Destruction: Culture and Mass Killing in the First World War*. New York: Oxford University Press, 2007.

Laffin, John. *British Butchers and Bunglers of World War One*. Gloucester: Alan Sutton, 1988.

Lee, John. "Some Lessons of the Somme: The British Infantry in 1917," *Look to your front, Studies in the First World War*. Brian Bond et al. Staplehurst: Spellmount, 1999.

Lengel, Edward G. *To Conquer Hell: The Meuse-Argonne, 1918 The Epic Battle That Ended the First World War*. New York: Lippincott Williams & Wilkins, 2009.

Livesay, J. F. B. *Canada's Hundred Days: With the Canadian Corps from Amiens to Mons, Aug. 8 - Nov. 11, 1918*. Toronto: Thomas Allen, 1919.

Luciuk, Y. Lubomyr and Soroby, Ron. *Konowal: A Canadian Hero*. Kingston: Kashtan Press, 2000.

Lupfer, Timothy T. *Leavenworth Papers, No. 4, Dynamics of Doctrine: The Changes in German Tactical Doctrine During the First World War*. Leavenworth: U.S. Army Command and General Staff College, 1981.

Lynch, John William. *Princess Patricia's Canadian Light Infantry, 1917-1919*. Hicksville: Exposition Press, 1976.

Macdonald, Lyn. *1915: The Death of Innocence*. Baltimore: Johns Hopkins University Press, 2000.

MacGowan, S. Douglas; Heckbert, Harry M.; and O'Leary, Byron E. *New Brunswick's 'Fighting 26th': A History of the 26th New Brunswick Battalion, C.E.F. 1914-1919*. Sackville: Neptune Publishing, 1994.

Machum, George C. *The Story of the 64th Battalion, C.E.F.: 1915-1916*. Montreal: Industrial Shops for the Deaf, 1956.

Marshall, Samuel Lyman Atwood. *World War I*. Boston: Houghton Mifflin Harcourt, 2001.

Martin, Stuart. *The Story of the 13th Battalion, 1914-1917*. London: Canadian War Records Office, 1917.

Matheson, William D. *My Grandfather's War: Canadians Remember the First World War 1914-1918*. Toronto: Macmillan of Canada, 1981.

McCarthy, Chris. "Queen of the Battlefield: The Development of Command, Organization and Tactics in the British Infantry Battalion during the Great War." *Command and Control on the Western Front: The British Army's Experience, 1914-1918*. ed. Sheiffeild, Gary and Todman, Dan. Staplehurst: Spellmount, 2004.

McClintock, Alexander. *Best O' Luck: How a Fighting Kentuckian won the Thanks of Britain's King*. Vernon: CEF Books, 2000.

McEvoy, Bernard and Finlay, A. H. *History of the 72nd Canadian Infantry Battalion: Seaforth Highlanders of Canada*. Vancouver: Cowan and Brookhouse, 1920.

McKean, G. B. *Scouting Thrills*. New York: MacMillan Company, 1919.

McMillan, David. *Trench Tea and Sandbags*. Canada: R. McAdam, 1996.

McWilliams, James L. and Steel, R. James. *The Suicide Battalion*. St. Catharines: Vanwell Publishing, 1990.

Millar, W. C. *From Thunder Bay Through Ypres with the Fighting 52nd*. Canada: s.n., 1918.

Ministry Overseas Military Forces of Canada. *Report of the Overseas Military Forces of Canada, 1918*. London: 1918.

Mitchell, T. J. and Smith, G. M. *Medical Services: Casualties and Medical Statistics of the Great War*. London: His Majesty's Stationary Office, 1931.

Morrison, James C. *Hell on Earth, A personal account of Prince Edward Island soldiers in The Great War*. Summerside: J. C. Morrison 1995.

Morrow, John Howard and Morrow, J. Jr. *The Great War: An Imperial History*. London: Routledge, 2005.

Morton, Desmond. *When Your Number's Up: The Canadian Soldier in the First World War*. Toronto: Random House of Canada, 1993.

Morton, Desmond and Granatstein, J. L. *Marching to Armageddon: Canadians and the Great War 1914-1919*. Toronto: Lester & Orpen Dennys, 1988.

Murray, W. W. *The History of the 2nd Canadian Battalion (East. Ontario Regiment) Canadian Expeditionary Force in the Great War 1914-1919*. Ottawa: Mortimer, 1947.

Nickerson, Hoffman. *The Armed Horde*. New York, 1940

Nicholson, G. W. L. *Canadian Expeditionary Force, 1914-1919*. Ottawa: Queen's Printer and Controller of Stationary, 1964.
    ---. *The Gunners of Canada, The History of the Royal Regiment of Canadian Artillery, Volume 1, 1534-1919*. Toronto: McClelland and Stewart, 1967.

Palazzo, Albert *Seeking Victory on the Western Front: The British Army and Chemical Warfare in World War 1*. Lincoln: University of Nebraska Press, 2000.

Peat, Harold R. *Private Peat*. Indianapolis: Bobbs-Merrill Co., 1917.

Radley, Kenneth. *We Lead Others Follow: First Canadian Division, 1914-1918*. St. Catharines: Vanwell, 2006.

Ramsay, M. A. *Command and Cohesion, The Citizen Soldier and Minor Tactics in the British Army, 1870-1918*. Westport: Praeger, 2002.

Rawling, Bill. *Surviving Trench Warfare: Technology and the Canadian Corps 1914-1918*. Toronto: University of Toronto Press, 1992.

Reid, Gordon. *Poor Bloody Murder: Personal Memoirs of the First World War*. Oakville: Mosaic Press, 1980.

Ripley, Tim. *Bayonet Battle, Bayonet Warfare in the 20th Century*. London: Sidgwick & Jackson, 2002.

Rubin, Louis Decimus. *The Summer the Archduke Died: On Wars and Warriors*. (University of Missouri Press, 2008).

Russenholt, E. S. *Six Thousand Canadian Men: Being the History of the 44th Battalion Canadian Infantry 1914-1919*. Winnipeg: De Montford Press, 1932.

Samuels, Martin. *Command or Control? Command, Training and Tactics in the British and German Armies, 1888-1918*. Portland: Frank Cass, 1995.

Sanders, Charles W. Jr. *No Other Law: The French Army and the Doctrine of the Offensive*. Santa Monica: Rand Corporation, 1987.

Shackleton, Kevin R. *Second to None: The Fighting 58th Battalion of the Canadian Expeditionary Force*. Toronto: Dundurn Group, 2002.

Sheffield, Gary. *Forgotten Victory, The First World War: Myths and Realities*. London: Headline, 2001.
    ---. "Vimy Ridge and the Battle of Arras: A British Perspective." *Vimy Ridge, A Canadian Reassessment*. Ed. Geoffery Hayes, Andrew Iarocci, and Mike Bechthold, Waterloo: Wilfrid Laurier University Press, 2007.

Sheldon, Jack. *The German Army on the Somme 1914-1916*. Barnsley: Pen & Sword, 2006.
    ---. *The German Army on at Passchendaele*. Barnsley: Pen & Sword, 2007.
    ---. *The German Army on Vimy Ridge 1914-1917*. Barnsley: Pen & Sword, 2008.
    ---. *The German Army at Cambrai*. Barnsley: Pen & Sword, 2009.

Showalter, Dennis E. "Prussia, Technology and War: Artillery from 1815 to 1914," *Men, machines & war*. Ed. Ronald Haycock and Keith Neilson. Waterloo: Wilfred Laurier University Press, 1988).

Singer, H. C. *History of the 31st Canadian Infantry Battalion C.E.F.* Calgary: Detselig, 2006.

Smith, Leonard V.; Audoin-Rouzeau, Stéphane; Becker, Annette. *France and the Great War, 1914-1918*. New York: Cambridge University Press, 2003.

Stevenson, David. *Cataclysm, The First World War as Political Tragedy*. New York: Basic Books, 2004.

Stirrett, George. *A Soldier's Story – 1914-1918*. Unpublished manuscript, Canadian War Museum.

Stone, Jay and Schmidl, Erwin A. *The Boer War and Military Reforms*. Lanham: University Press of America, 1988.

Strachan, Hew. *The First World War, Volume 1: To Arms*. New York: Oxford University Press, 2001.

Taylor, A. J. P. *Illustrated History of the First World War*. New York: G. P. Putnam's Sons, 1964.

Tascona, Bruce and Wells, Eric. *Little Black Devils: A History of the Royal Winnipeg Rifles*. Winnipeg: Frye Publishing, 1983.

Travers, Tim. *The Killing Ground: The British Army, the Western Front and the Emergence of Modern Warfare 1900-1918*. London: Allen & Unwin, 1987.

Terraine, John. *The Great War*. London: Wordsworth Editions, 1997.

The Adjutant. *The 116th Battalion in France*. Toronto: E. P. S. Allen, 1921.

Todman, Dan. *The Great War, Myth and Memory*. New York: Hambledon, 2005.
   ---. "The Grand Lamasery revisited: General Headquarters on the Western Front, 1914-1918." *Command and Control on the Western Front: The British Army's Experience, 1914-1918*. ed. Gary Sheiffeild and Dan Todman, Staplehurst: Spellmount, 2004.

Todman, Dan and Sheffield, Gary. "Command and Control in the British Army on the Western Front." *Command and Control on the Western Front: The British Army's Experience, 1914-1918*. ed. Gary Sheiffeild and Dan Todman, Staplehurst: Spellmount, 2004.

Tooley, Hunt. *The Western Front, Battle Ground and Home Front in the First World War.* New York: Palgrave, 2003.

Topp, C. Beresford, *The 42nd Battalion, C.E.F. Royal Highlanders of Canada in The Great War.* Montreal: Gazette Printing, 1931.

Urquhart, H. M. *The History of the 16th Battalion (The Canadian Scottish) Canadian Expeditionary Force in the Great War, 1914-1919.* Toronto: MacMillan, 1932.

Hodder-Williams, Ralph. *Princess Patricia's Canadian Light Infantry, 1914-1919.* Toronto: Hodder and Stroughton, 1923.

Winter, Denis. *Death's Men: Soldiers of the Great War.* Markham: Penguin Books, 1979.

Winter, Jay M. *The Experience of World War I.* New York: Oxford University Press, 1989.

Winter, Jay M. and Baggett, Blaine. *The Great War And the Shaping of the 20th Century.* New York: Penguin Studio, 1996.

Woods, Timothy; Wiest, Andrew A.; and Barbier, M. K. *Strategy and Tactics, Infantry Warfare.* St. Paul: Zenith Imprint, 2002.

Vance, Jonathan F. *Death So Noble: Memory, Meaning, and the First World War.* Vancouver: UBC Press, 1999.

Various Authorities. *Canada in the Great World War: An authentic account of the Military History of Canada from the earliest days to the close of the war of the Nations. Volume VI: Special Services, Heroic Deeds, Etc.* Toronto: United Publishers of Canada.

Zuber, Terence. *The Battle of the Frontiers: Ardennes 1914.* Chalford: Tempus, 2007.

Journal and Periodical Articles

Adair, Jason. "The Battle of Passchendaele, The experiences of Lieutenant Tom Rutherford, 4th Battalion, Canadian Mounted Rifles." *Canadian Military History* 13, No. 4, Autumn 2004.

Arnold, Joseph C. "French Tactical Doctrine, 1870-1914." *Military Affairs* 42, No. 2, (April 1978).

Brown, Ian H. "Not Glamorous, But Effective: The Canadian Corps and the Set-piece Attack, 1917-1918." *The Journal of Military History* 58, (July 1994).

Chappelle, Dean. "The Canadian Attack at Amiens, 8-11 August 1918." *Canadian Military History* 2, no. 2, (1993).

Cook, Tim. "The Blind Leading the Blind, The Battle of the St. Eloi Craters." *Canadian Military History* 5, No. 2, (Autumn 1996), 34.
---. "The Politics of Surrender, Canadian Soldiers and the Killing of Prisoners in the Great War." *Journal of Military History*. 70 (July 2006).

Cox, Gary J. "Of Aphorisms, Lessons, and Paradigms: Comparing the British and German Official Histories of the Russo-Japanese War." *The Journal of Military History* 56. (April 1992).

Engen, Rob. "Steel against Fire: The bayonet in the First World War." *Journal of Military and Strategic Studies* 8, No. 3, (Spring 2006).

Echevarria II, Antulio J. "The 'Cult of the Offensive' Revisited: Confronting Technological Change Before the Great War." *Journal of Strategic Studies*. Vol. 23, No. 1, (March 2002), 201.

Ferguson, Niall. "Prisoner Taking and Prisoner Killing in the Age of Total War: Toward a Political Economy of Military Defeat." *War in History* 11, No. 2 (April 2004).

Grodzinski, John. "The Use and Abuse of the Battle: Vimy Ridge and the Great War over the History of the First World War." *Canadian Military Journal,* Vol. 10, No. 1.

Guthrie, P. A. "Festubert: A Graphic Story of a Great Fight Where Canadians Won Honor at a Heavy Price Just a Year Ago." *The Montreal Daily Star*. (May 27th, 1916).

Harvey, A. D. "The Bayonet in Battle." *RUSI Journal.* Vol. 150, No. 2, (April 2005).

Haynes, Alex D. "The Development of Infantry Doctrine in the Canadian Expeditionary Force." *Canadian Military Journal*. (Autumn 2007).

Hodges, Paul. "'They don't like it up 'em': Bayonet fetishization in the British Army during the First World War." *Journal of War and Culture Studies*. Volume 1, Number 2.

Howard, Michael. "Men Against Fire: Expectation of War in 1914." *International Security*, Vol. 9, No 1. (Spring 1984).

Humphries, Marc Osborne. "The Myth of the Learning Curve, Tactics and Training in the 12th Canadian Infantry Brigade, 1916-1918." *Canadian Military History* 14, No. 4 (Autumn 2005).

Iarocci, Andrew. "1st Canadian Infantry Brigade in the Second Battle of Ypres, The Case of the 1st and 4th Canadian Infantry Battalions, 23 April 1915." *Canadian Military History* 12, Number 4, (Autumn 2003).

Kent, Roland G. "The Military Tactics of Caesar and To-Day." *The Classical Weekly* 8, No. 9 (Dec. 12, 1914).

McGuffie, T. H. "The Bayonet: A survey of the weapon's employment in warfare over the past three centuries." *History Today* 12 (August 1962).

Morton, Desmond. "Changing operational Doctrine in the Canadian Corps 1916-1917." *The Army Doctrine and Training Bulletin* 2, No. 4, (Winter 1999).

O'Leary, M. M. "À la bayonet or Hot Blood and Cold Steel." *Canadian Army Infantry Journal*. (Spring 2000).

Seki, T. "The Value of the *Arme Blanche*, with illustrations from the recent Campaign." Trans. F. S. G. Piggott. *Royal United Service Institute Journal.* Vol. 55, part 2 (July-Dec. 1911).

Setzen, Joel A. "Background to the French Failures of August 1914: Civilian and Military Dimensions." *Military Affairs* 42, No. 2, (April 1978).

Sorobey, Ron. "Filip Konowal, VC, The Rebirth of a Canadian Hero." *Canadian Military History*. Vol. 5, No, 2, (Autumn 1996).

Travers, Tim. "The Offensive and the Problem of Innovation in British Military Thought 1870-1915." *Journal of Contemporary History* 13 (1978).

-----. "Learning and Decision-Making on the Western Front, 1915-1916: The British Example." *Canadian Journal of History* 18, No. 1, (April 1983).

Todd, Frederick P. "The Knife and Club in Trench Warfare, 1914-1918." *The Journal of the American Military History Foundation* 2, No. 3, (Autumn 1938).

Theses and Dissertations

Campbell, David Charles Gregory. "The Divisional Experience in the C.E.F.: A Social and Operational History of the 2nd Canadian Division, 1915-1918." PhD dissertation: University of Calgary, 2003.

Campbell, James Dunbar "The Army isn't All Work: Physical Culture and the Evolution of the British Army 1860-1918." PhD dissertation: University of Maine, 2003.

Cook, Tim. "No Place to Run." M.A. Thesis: Royal Military College of Canada, 1997.

Hamric, Jacob Lee. "Germany's Decisive Victory: Falkenhayn's Campaign in Romania, 1916." M.A. Thesis: Eastern Michigan University, 2004.

Heidt, Daniel "From Bayonets to Stilettos to UN Resolutions: The Development of Howard Green's Views Regarding War." M.A. Thesis: University of Waterloo, 2008.

Murray, Nicholas A. A. "The Theory and Practice of Field Fortification from 1877-1914." P.h.D. Dissertation: St. Anthony's College, University of Oxford, 2007.

Stewart, William Frederick. "Attack Doctrine in the Canadian Corps, 1916-1918." M.A. Thesis: University of New Brunswick, 1982.

Index

Abatis. 60
Alley, H. R. 80, 91
Amiens. .... vi, 4, 47, 53, 56, 135, 136, 138, 140, 146
Ancre Heights. 98
Anderson, Percival W. 114, 115, 129
Angelo, Charles Henry. 21, 36, 38, 44
Arleux. 115
    attack at. 115
Armagh Wood. 91
Arras. 3, 4, 53, 115, 116, 140, 143, 146
Arras offensive (1917). 115, 116
Arras, battle of (1918). 4, 140, 143, 145, 146
artillery. 1, 4, 13, 14, 23-26, 28, 30, 35, 55, 56, 58, 60, 61, 69-71, 73-76, 84-94, 96, 98, 99, 101, 104, 109-113, 115, 117, 124, 128, 137, 146
    ammunition. 87, 92, 93, 98
    counter-battery fire. 112, 114, 124, 150
    creeping barrage. 93, 95, 98, 101, 111, 112, 114, 117, 122, 124, 136, 144, 145, 150
    feint. 89
    fuses. 87, 88, 93, 98
    observed fire. 60
    predicted fire. 60, 114
    standing barrage. 112
    taken with bayonet charge. 137
    worn gun barrels. 104

assault course. 48, 50, 63, 78
assault training. 120, 121, 131, 132, 142
Assault Training 1917. 120, 121, 131, 132, 142
Bagshaw, F. G. 79
Bairnsfather, Bruce. .... v, 12, 15
Balck, Hermann. 31
Baldwin, Harold. 16, 51, 74, 80
barbed wire. 4, 18, 60, 87, 96, 140
    cutting. 96
bayonet. ..... 2, v, 1-11, 13-60, 62-68, 70, 72-99, 102-112, 114, 115, 117, 119-131, 133-140, 142-151
    charge. 2, 8-10, 13, 14, 23-26, 35, 39, 40, 49, 50, 52-58, 64, 65, 67, 68, 73, 87, 91, 96, 102, 103, 114, 119, 121, 128-131, 137, 138, 143-146, 148-151
    counter-charge. 73, 120, 122
    cult of. 24
    misuses of. 11, 15
    moral revulsion to. 11, 20, 21
    myth of obsolescence. 1, 2, 5, 9-11, 23, 26
bayonet fighting. ...... v, 2-6, 9-11, 13, 15, 19-22, 24, 25, 28, 34, 36-42, 44-51, 54, 56, 57, 62, 63, 65, 68, 77, 79-84, 89, 95, 97, 98, 104-109, 114, 115, 117, 119, 122, 123, 129-131, 133, 135, 136, 140, 143, 146-149, 151

Bayonet Fighting for Platoon Commanders 25, 54, 57
Bayonet Fighting Illustrated 1917...... v, 19, 20, 46, 51, 80, 131, 143
bayonet fighting schools.......... 82, 104
    Calgary.............. 104
    Halifax. 104
    London.............. 104
    Montreal............ 82
    Ottawa............... 82
    Shorncliffe. 82, 104, 105
    Toronto. 104
    Winnipeg............ 82
bayonet fighting, techniques of
    butt stroke. 78, 84, 107
    duck. 46, 131
    feint.................. 46, 107
    firing a round. 80, 143
    infighting...... 5, 40-42, 44, 47, 78, 81, 83, 107, 131, 133, 142
    jab..................... 44, 49, 80, 83, 131, 142
    killing face. 57, 78, 81
    knee to the groin.......... 47, 143
    long point. 39, 40, 44, 47, 78, 107
    parry......... 37-39, 45, 46, 63, 78, 84, 107, 142
    planting the foot. 80, 143
    shorten arms............. 40, 41, 44, 47, 142
    tripping................ 41
    twist. 80, 81, 105, 108, 143
bayonet instructors. 62, 105
    AGS................. 81, 82
    assistant............. 62, 81, 82, 105
    Imperial............. 62
    private. 36, 38
    unofficial techniques used by. 77, 81, 105
bayonet training........ 2-5, 10, 13, 20-22, 37, 38, 44, 46-49, 51, 52, 57, 62, 63, 77, 79-82, 84, 104, 105, 107, 130, 131, 135, 142, 143, 150
    assault course. 48, 50, 63, 78
    tools................. 108
Bayonet Training 1916... 13, 21, 22, 46, 47, 51, 80, 84, 104, 107, 143, 200
Beaucourt Wood. 137, 138
Bell, George V.. .... 44, 51, 91, 92
Bidwell, Shelford. 1, 29, 35, 60, 62, 92, 93, 109
Bird, William. 45, 47, 48, 55, 56, 130, 136, 143, 147
blobstick................ v, 63, 106-108, 131
Boer War. 1, 28, 29, 31-35, 38, 57, 60
Boguslawski, Albrecht von............ 31
bomb........ 18, 47, 51, 64, 74-76, 87, 93-95, 98, 110, 111, 116, 124, 127, 131-133, 141

bombing.... 49, 63-65, 73, 75, 87, 91, 102, 103, 116, 123, 126, 135
bounding. 24, 28, 35, 66-68, 70, 114, 117, 119, 132, 138, 143, 144
Brazier, The............ 16, 79
British Imperial Army............ 61
    7th Division............ 68
    Army Gymnastic Staff (AGS). 37, 81, 82, 105
    British Expeditionary Force (BEF). 3, 61
    British First Army................ 116, 124
    doctrine. 3, 26
Burton, Sir Richard............ 36-38, 44
Cagnicourt. 145, 146
Cambrai. 124, 136, 146, 147
Campbell, Ronald......... 37, 57, 61, 79, 88, 112, 116, 129, 134
Canadian Expeditionary Force......... 2, 9, 13, 29
    102nd Battalion. 6, 137, 138
    10th Battalion......... 53, 66, 67, 70-73, 120
    116th Battalion............ 136
    13th Battalion............. 66, 117, 120, 145
    14th Battalion....... 15, 113, 114, 144, 145
    15th Battalion............ 70, 119
    16th Battalion........ 66, 67, 70, 90, 91, 93, 101, 110, 113, 120
    18th Battalion............ 54, 95
    1st Battalion............ 75, 89, 97
    1st Brigade. 25
    1st Canadian Division. 3
    1st Canadian Mounted Rifles (CMR) 89, 97, 140
    1st Canadian Mounter Rifles (CMR).... 99
    20th Battalion............ 121
    21st Battalion............ 95, 96, 151
    22nd Battalion............ 97
    25th Battalion............ 97, 122
    28th Battalion............ 95, 96
    29th Battalion............ 63, 122
    2nd Canadian Division. 88, 134
    2nd Canadian Mounter Rifles (CMR) 99, 140
    31st Battalion............ 95, 117
    3rd Battalion............ 74, 89, 91, 92, 103
    3rd Brigade. 66
    3rd Canadian Division........... 88
    42nd Battalion........... 45, 53, 136-138
    44th Battalion............ 65
    46th Battalion............ 47, 125
    47th Battalion............ 18
    49th Battalion............ 116
    4th Battalion............ 74, 119
    4th Brigade. 96
    4th Canadian Division........... 145
    4th Canadian Mounted Rifles (CMR) 89, 97, 127, 140
    58th Battalion............ 89

Index

5th Battalion. 51, 73, 79
5th Canadian Mounted Rifles (CMR)
.... 97, 99, 140
6th Brigade. 122
78th Battalion. 136, 138
7th Battalion. 73, 119
7th Brigade. 97, 140
85th Battalion. 14, 105, 114, 128-130
8th Brigade. 89, 97, 140
as CEF. 2, 49, 53, 54, 58, 105
Canadian Army Gymnastic Staff (CAGS)
.... 82, 105
Canadian Corps. 4, 5, 9, 13, 14, 53, 62,
68, 88, 89, 92, 94, 98, 105, 108, 112,
113, 116, 117, 122, 124, 128, 131-
135, 143, 146-148
First Contingent. 62, 80
Motor Machine Gun Brigade. 134
Princess Patricia's Canadian Light Infantry
(PPCLI). 61, 89, 140, 141
Royal Canadian Regiment (RCR). 57,
62, 101-103, 140, 141
Canal du Nord. 146, 147
cartography. 75
casualty statistics. 11, 97
fatalities. 13, 14, 145
ratios. 11, 13, 14, 145
wounds. 11, 13, 14
Chabelle, Joseph. 17, 97, 98
Clark, Alan. 8, 16
close combat. .. 1-3, v, 2, 3, 5, 6, 11, 12, 14, 17-
19, 21, 22, 26, 43, 44, 46-53, 55-58,
73, 74, 86, 91, 92, 96, 103, 117, 119,
120, 127, 130, 131, 133, 139, 143-
145, 148-151
negotiation of. 52, 55-57, 73, 86, 92,
103, 143, 149
stress of. 22, 47, 52, 55, 56
close with the enemy. 23, 25, 29, 49, 93, 113,
138, 151
cognitive awareness. 43
Committee on Imperial Defense. 3
conditioning. ... 10, 11, 43, 48-53, 55, 143, 144,
146
dehumanization. 10, 49, 50, 52
hate. 10, 11, 49
Cook, Tim. .... iii, 1, 5, 9, 11, 14, 16, 17, 23, 24,
33, 52, 54, 55, 58, 60, 62, 68, 69,
87-90, 92, 95, 97-99, 110, 113,
114, 117, 134, 135, 147
counter attack. 88, 103
cover. ... 24-26, 28, 29, 31-33, 56, 87, 110, 125,
139, 141
fire. 111, 125, 128, 138
night. 68, 71, 90, 92, 140
Cox, Sid. 8, 33, 53, 67
Crowe, Jordan. 17, 47, 136
cult of the offensive. 31

Currie, Sir Arthur. 110-112, 116, 117, 132, 134
Dalby, G.. .... 149-151, 198
Davidson, E.F.. 114
defence in depth. 115
dehumanization. 10, 49, 50, 52
Dinesen, Thomas. 47, 50, 51, 107, 136
Directorate of Bayonet Fighting and Physical
Training. 105, 143
disassociation. 47, 48, 52
dispersion. ... 27-36, 66, 84, 110, 111, 115, 132,
143
doctrine. 1-4, 8, 9, 23-26, 28, 29, 31, 33-35,
52, 53, 55, 57, 61, 87, 92, 115, 117,
136, 140
Doolan, William Patrick. 151
Douai Plain. 115
Dragomirov, Mikail. 31
Du Castelnau, Noel. 34
Du Picq, Charles Ardant. 31
Ecole Superieure du Guerre. 30
élan. 3, 27-36, 38, 39, 41, 75, 84, 121, 122,
150
Elgin Commission. 33, 36, 38
Ellis, John. 8, 11, 203
English Channel. 59
Ericson, Oscar. 57, 139
Fabeck Graben. 97, 99
fencing. .. 21, 22, 36-39, 44-46, 48, 62, 81, 107,
131
distance. 21, 44-46
early modern. 44
measure. 44, 46
theory. 44
time. 44-46
Ferdinand, Archduke Franz. 34, 59
Festubert. .... vi, 3, 68-71, 73, 74, 84, 88, 93
18 May attack. 68
20 May attack. 70
21 May attack. 71, 72
24 May attack. 73, 74
battle of. 3, 68, 73, 74, 84
fetish. 10, 49
fire and movement. 28, 30, 31, 34, 35, 90,
109-111, 114, 124, 125, 128, 132,
133, 137, 139
Foch, Ferdinand. 31, 34
French Army. 23, 28-31, 34
45th Algerian Division. 65
doctrine. 24, 31
Fresnoy, attack at. 115
Frontiers, battle of. 24, 34, 59, 211
Fuel Trench. 141
Fuller, G. A.. .... 2, 121
gallows sack. 39, 63
General Staff. 34, 55, 82, 92, 110, 132
German Army. .. 29, 32-34, 50, 55, 88, 95, 113-
115, 124
defensive tactics. .. 55, 115, 116, 132, 140

doctrine. 57, 140
German spring offensives. 134, 135
Gibbs, Stormont. 10, 11
Gilchrist, Harry L.. 11
Givenchy. .... vi, 3, 68, 75, 76, 85
Golden, Leo. 16, 73
Goodmurphy, Arthur. 55, 96, 136
Graham, Dominick..... 1, 29, 33, 35, 55, 57, 60,
62, 92, 93, 109
Grandmaisson, Francois Loyzeau de. 30, 31, 34
Grant, Enos. 44, 50, 51
Green, H. Arnson. 102
Green, Howard Charles. 49
grenades. 61, 64, 75, 84, 85, 110, 114, 116,
119, 125, 126, 133
hand. ... 3, 56, 61, 63, 74, 85, 87, 116, 119
rifle. 119, 125, 126, 128, 129, 133
tactics. 61
Griffith, Paddy... 7, 25, 31, 61, 87, 88, 93, 116,
127
groin, knee to. 20, 47, 131, 142, 143
Grossman, David. 11, 20, 21, 52
Guthrie, Percy. 70-72
Haig, Sir Douglas. 31
Hamon Wood. 136
hand to hand fighting. 89
Hanna, Robert. 123, 124
Hans Trench. 144
Harris, John. 3, 8, 128
Hill 102. .... 136-138, 149
Hill 145. 114
Hill 70. .... vi, 4, 53, 117-124, 149
Hindenburg Line. 143, 145
historiography. 2, 9, 30
history... 2, 1, 2, 5, 7-11, 13, 15, 17, 19, 20, 22,
34, 46, 48-50, 52, 53, 55-57, 61-70,
79, 82, 89-92, 95, 96, 98, 101, 104,
110, 113, 114, 116, 117, 121, 122,
125, 127, 128, 130, 135-138, 143,
147
Canadian historiography. 9
Hobson, Frederick. 121, 122
Hodges, Paul. 1, 10, 11, 17, 49, 50, 52
Hughes, Sir Sam. 62
Hurley, J.F.. 114
Hutton, Alfred. 36, 38, 44
In Flanders Fields. 6, 54, 116, 135, 151
infantry regulations, British
1902. .... 24, 35
1905. .... 38, 39
infantry regulations, French
1875. 31
1884. 31
1904. .... 33, 34
1913. .... 34, 35
infantry regulations, German
1873. 32
1888. 32-34

1906. 34
Infantry Training 1902. .... 24, 38, 39
Infantry Training 1905. .... 24, 38-40
Infantry Training 1911.... 22, 24, 25, 39, 40, 83
Infantry Training 1914..... 2, 22, 24, 25, 62, 78,
110, 125
infighting. . 40-42, 44, 47, 78, 81, 83, 107, 131,
133, 142
initiative. 27, 29, 30, 33, 50, 109-112, 116,
122, 125, 128, 130, 132, 135, 139
of sub-alterns and NCOs. 29, 110, 116,
128, 135
instinct. 3, 21, 22, 43-49, 51, 55, 57, 72
self-preservation. 3, 11, 43, 46, 48, 49,
53, 55, 57
swing in defense. 22
instructors. 36, 38, 44, 49, 51, 62, 81, 82, 84,
104-106
intimidation. 56-58, 139
yell. 57, 70, 72, 119, 139
jab. 44, 83, 131, 142
Jigsaw Wood. 141
Joffre, Joseph. 31
K5. .... 70-73, 75
Kempling, George Hedley. 107
killing face. 57, 78
Kitcheners Wood. 5, 66-68
knob kerry. 17
learning curve. 9, 213
Lens. 117, 121, 122
Lewis gun. 84, 94, 110, 117, 119, 121, 129,
146
clubbed. 119
loose play. 39
Lunn, Charles. 5, 6, 66, 67
Lynch, John William. 80, 140-143
MacFarlane, W.C. 120, 145
machine gun. 11, 48, 53, 54, 66, 71, 74, 75,
85, 87, 90, 91, 101, 103, 110, 114,
119, 120, 122, 123, 125, 126, 128,
133, 134, 136, 138-142, 144-146
MacMillan, David. ... 1, 16, 56, 57, 81, 110, 122
Maheux, Frank. 96, 97
Marne. 59, 135
1st battle of. 59
2nd battle of. 135
Maud'huy, Louis. 29
Mauser Ridge. 65, 66, 68
Mayes, Henry Geroge. 82, 104, 105, 131
McGuffie, Tom Henderson. 11, 14, 15
McKean, George Burdon.. .... 57, 144
McLean, A.L. 145, 146
McLennan, A.M. 70, 71
Meckel, Jacob. 29
Memorandum of the Training and employment of Grenadiers. 64
Memorandum on trench to trench attacks. 25, 30

Methods of Unarmed Attack and Defense 130, 131
Milan Trench. .... vi, 141
military tradition. 1
mining.. 76, 85
Mitchell, T.J... 11
mobilization. 3, 5, 22, 62
Morrison, Corporal................................... 49, 53
mortar. 61, 84, 138
    Cohorn. 61
    Stokes. 111
Morton, Desmond. . 2, 9, 11, 17, 24, 49, 60, 62, 87-89, 112
Mount Sorrell
    battle of 13 June. 56, 89
    battle of 2 and 3 June. 88, 89, 92
Mouquet Farm. 97
Nabob Alley. 121
narrative....... 7-9, 16, 26, 53, 71, 130, 136, 140, 143, 144, 146, 147
    "learning curve". 9, 26
    "lions and donkeys"............ 7-9, 14, 16, 24
negotiation of close combat..... 3, 52, 55-57, 73, 86, 92, 103, 143, 149
negotiation of surrender............. 14, 53, 54, 146
Nicholson, William. . 31, 62, 65, 89, 92, 93, 96, 99, 101, 104, 112, 114, 115, 117, 121, 136, 143
no man's land.... 1, 4, 24, 26, 49, 68, 74-76, 86-88, 98, 99, 122, 148, 149
Nun's Alley...................................... vi, 122, 123
objective lines. 94
officers........... 8, 17, 29, 33, 35, 37, 56, 66, 103, 105, 110, 114, 116, 127, 129, 144
    devolution to subalterns........ 29, 110, 112, 129, 130
    shortage of.. 29
Oh What a Lovely War.................................... 8
Orange Hill. 140
Organization of Bayonet Fighting and Physical Training................................ 13, 106
Palazzo, Albert.......................... 7, 35, 88, 111
Passchendaele. ..... vi, 4, 8, 45, 53, 54, 115, 124-127, 130, 134
    26 October battle of.................... 125, 127
    30 October battle of........................... 128
Pearkes, George. 98, 101, 135
Peat, Harold. 16
Pelves... 141
Perman, Dan. 53
Peronne Wood... 137
personal accounts. 2, 5, 6, 16, 17, 148, 149
Petain, Phillippe. 23, 31
pillbox... 125, 127
Plan XVII.. 23, 59
platoon. 25, 51, 53, 54, 56, 57, 93, 105, 109-112, 114, 116, 117, 120, 129, 130, 133, 135, 137, 138, 146

emphasis on in the attack..... 93, 112, 116, 133
playing dead.. 52, 55
popular culture. 1, 7, 8
Port Arthur... 33
Practical Bayonet Fighting. 77, 81, 82, 143
prisoners. 6, 10, 14, 52, 53, 58, 97, 102, 113, 119, 137-139, 146, 147
    killing of................................... 10, 52, 53
    taking of.. 53, 58, 97, 102, 113, 119, 137-139, 146, 147
Radcliff, Percy. 110
Ramsay, M.A....... 24-26, 28, 29, 33, 61, 62, 92, 111
Rawling, Bill.... 1, 9, 11, 18, 24, 32, 33, 60, 61, 70, 76, 87, 88, 92, 93, 99, 101, 111-116, 121, 133, 138
records. ... 2, 5, 6, 13, 53, 57, 62, 66, 73, 74, 89, 91, 92, 95, 114, 119, 122, 129, 130, 136, 140, 141, 146-149
    ambiguous language of..................... 141
    battalion histories. ... 2, 5, 6, 9, 15, 66, 70, 90, 110, 113, 114, 117, 128, 136, 137, 148, 149
    official.......... 2, 5, 6, 13, 73, 95, 122, 129, 136, 141, 148, 149
    personal... 6
Regina Trench............................. vi, 4, 99-104
    1 October attack on. 99
    8 October attack on. 101, 104
rifles..... 3, 17, 18, 20-22, 25, 28, 32, 40, 44, 47, 48, 51, 52, 56, 57, 62, 66, 74, 75, 77, 79-81, 83, 84, 90-92, 94, 95, 98, 99, 107, 110, 111, 114, 116, 117, 119-121, 124-129, 131-133, 142, 143, 146, 149, 150
    Chassepot.. 28
    Dreyse needle gun. 28
    Lee-Enfield............................. 16, 79, 80
    Ross. 62, 65, 79, 80
Ripley, Tim. 10, 33
rushes.......... 24, 25, 70, 111, 128, 132, 137-139, 142, 144
Rutherford, Tom. 127
S.S. 143 Instructions for the Training............ 51
sack, bayonet................................... 39, 63, 82
Sarajevo... 59
Saunders, Charles....................................... 23
Scarpe............................. 4, 13, 140, 144-146
    battle of Milan Trench........................ 141
    battle of the. 4, 146
    Fuel Trench. 141
Schlieffen Plan....................................... 34, 59
Second World War............................. 7, 8, 135
Sedan, battle of. 28
self-preservation... 3, 11, 43, 46, 48, 49, 53, 55, 57
Setzen, Joel. 29

Index

shock tactics........... 2, 3, 21, 32, 34, 42, 60, 150
Silver, George. 44
Sinclair, A.G. 16
Smith, G.M. 11
Smith, Sid. 47, 95
snipers... 25, 79
Somme........ 1, 8, 11, 23, 25, 26, 53, 55, 57, 87,
  88, 92-94, 98, 99, 101, 103, 109,
  110, 114, 116, 124, 129, 135
  battle of. 8, 25, 26, 55, 87, 92-94, 109,
    110, 116, 129
  first day of. 8, 23, 93, 94
Souchez. 14, 115
South Africa.. 32, 33
spirit of the bayonet.. 51
spirit of the offensive. 29
St. Eloi Craters........................... 88, 89, 92, 212
St. Patrick. 18
St. Privat. 28
sugar factory. 96, 97
surrender..... 3, 4, 10, 14, 43, 45, 49, 52-55, 58,
  91, 92, 96, 113, 114, 117, 133, 144,
  146, 148-151
tactics... 1-5, 7, 9, 10, 21, 24, 25, 27-36, 38, 41,
  42, 55, 57, 60, 61, 63, 64, 66-68, 75,
  76, 84, 86-88, 90, 92-94, 98, 109-
  111, 115-117, 119, 124, 125, 127-
  130, 132, 133, 136, 138, 139, 143,
  146, 150
  British. 10, 32, 93
  combined arms.................................... 25
  elastic defence..................................... 115
  German. 57
  set-piece battle.............. 2, 56, 75, 86, 109
tanks. 4, 95-97, 136-138, 140, 150
Tate, Edward.. 138, 139
technology....... 1, 2, 4, 8, 23, 26-28, 42, 60, 84,
  94, 98, 109, 112, 138
  new. 27, 28, 94, 109, 112, 137
Territorial (Reserve) Force. 36
Theipeval. 99
Thompson, F. G. 136
Thompson, J. H... 95
training......... v, 1-5, 8-10, 13, 17-26, 28-30, 32,
  33, 36-40, 42-57, 61-67, 77-85, 90,
  92, 94, 95, 104-108, 110, 111, 116,
  119-122, 124, 125, 130-135, 138,
  142, 143, 150, 151
  bayonet...... 2-5, 10, 22, 37, 38, 44, 46-49,
    51, 52, 57, 62, 63, 77, 79-82, 84,
    104, 105, 107, 142, 143, 150
  infantry... 77, 94
  musketry. 32, 110, 116, 132, 135
  training center
    Aldershot...................................... 82, 105
    Le Havre. 56, 131
    Salisbury Plain. 51, 62
    Shorncliffe. 82, 104, 105

training literature...... 17, 20, 22, 23, 25, 44, 46,
  51, 54, 57, 77, 78, 80, 81, 107, 111,
  120, 143
  official........................ 20, 78, 80, 81, 143
  unofficial... 77
training systems. 5, 62, 84, 105, 143
trench..... vi, 1, 3, 4, 8, 9, 11, 15-20, 22, 24, 25,
  30, 32, 33, 51, 56-61, 63-65, 67, 69-
  76, 79-81, 84, 86-88, 91-95, 99-104,
  111-116, 119-122, 124, 127, 132,
  133, 135, 138, 141-145
trench mortar... 138
trench storming parties........................... 18, 64
trench warfare
  problem of the offensive...................... 86
Valcartier. 3, 61, 62
Vance, Johnathan F......................... 9, 10, 211
Victoria Cross. 5, 13, 47, 121, 124, 139
Vienna Cottage.................................... 128, 129
Vimy Ridge................. vi, 3, 9, 49, 55, 109-115
wall bag... 63
Wilhelm II, Kaiser.. 34
Williams, J.S................................... 23, 87, 89
Winter, Denis... 1, 2, 11, 13, 14, 23, 24, 34, 49,
  51, 52, 61, 134
Yeo E.L... 64
Ypres. 3, 16, 51, 53, 59, 65, 68, 90, 117, 124
  1st battle of. 59
  2nd battle of. 65
  3rd battle of (see also Passchendaele)
    124
Zeddeler, L.L.. 31

About the Author

AARON MIEDEMA OFTEN seems to go where the wind will blow him. His path to this book was long and winding. He started out studying theatre, coming to serve as the Artistic Director of the Renaissance Stage Company in Kingston, Ontario, between 1992 and 2000. It was during his time in the theatre that he developed an interest in stage combat and martial arts, and after 2000 he began to dedicate himself to the study and practice of ancient and historical fencing. In 2005 he returned to Carleton University in Ottawa to pursue studies in Italian, Latin, and the social history of late medieval and early modern Europe. With this knowledge, Aaron began the translation and interpretation of Italian martial arts of the fifteenth, sixteenth, and seventeenth centuries.

However, at Carleton University Aaron's life took another peculiar turn – it was here that he met Tim Cook, who focused these interests into the study of the role of close combat in tactical doctrine on the Great War, which became the subject of his thesis for his Master of Arts in War Studies at the Royal Military College of Canada. He is now seeking a PhD program in order to become a professor of history – not surprisingly, the speciality has yet to be determined. He currently resides in Scarborough, Ontario, with his partner Yvonne.

Also Available from Legacy Books Press

The War that Changed the World
The Forgotten War that Set the Stage for the Global Conflicts of the 20th Century and Beyond

By John-Allen Price

ISBN: 978-0-9784652-1-6

Between 1870 and 1871, the world changed forever.

The Franco-Prussian War is often a forgotten war, its significance lost amidst larger conflicts such as the Napoleonic Wars and World War I. But, while it lasted less than a year, its aftermath would shape the course of history for decades to come.

In this comprehensive and epic account, John-Allen Price explores how this short but far-reaching war came to be, bringing the men who shaped history to life. Price examines the Franco-Prussian War and its world, from the seeds of the war in the Age of Napoleon to the Paris Commune, and the aftershocks that led to a century of slaughter, a war to end all wars, and an even greater war after that.

The Face of the Foe
Pitfalls and Perspectives of Military Intelligence

By Kjeld Hald Galster

ISBN: 978-0-9784652-6-1

Every nation that goes to war has to create images of their enemy. Through intelligence gathering and propaganda, these images are created and used to drive public support and keep soldiers fighting. At the same time, decision-makers must be provided with clear and incisive information on the opposition at hand. Frequently, these aims are mutually conflicting. Carefully balanced and used with circumspection, these images can lead to victory – but they can also drive armies to disaster and entire nations to atrocity.

In this sweeping and fascinating survey, Kjeld Hald Galster explores how intelligence is collected and interpreted. Drawing from examples ranging from the Napoleonic Wars to the 2003 War in Iraq, he examines how military intelligence is used to create the face of the foe – and what makes it a tremendous success...or a disastrous failure.